International Scientific and Research Conference on Knowledge-based Technologies in Development and Utilization of Mineral Resources 2016

IOP Conference Series: Earth and Environmental Science
Volume 45

Novokuznetsk, Russia
7 – 10 June 2016

Editor:

Viktor Fraynov

ISBN: 978-1-5108-3052-3
ISSN: 1755-1307

Printed from e-media with permission by:

Curran Associates, Inc.
57 Morehouse Lane
Red Hook, NY 12571

Some format issues inherent in the e-media version may also appear in this print version.

Printed by Curran Associates, Inc. (2016)

For permission requests, please contact the Institute of Physics
at the address below.

Institute of Physics
Dirac House, Temple Back
Bristol BS1 6BE UK

Phone: 44 1 17 929 7481
Fax: 44 1 17 920 0979

techtracking@iop.org

Additional copies of this publication are available from:

Curran Associates, Inc.
57 Morehouse Lane
Red Hook, NY 12571 USA
Phone: 845-758-0400
Fax: 845-758-2633
Email: curran@proceedings.com
Web: www.proceedings.com

Table of contents

Volume 45

International Scientific and Research Conference on Knowledge-based Technologies in Development and Utilization of Mineral Resources

7–10 June 2016, Novokuznetsk, Russian Federation

Accepted papers received: 23 September 2016
Published online: 4 October 2016

Preface

011001
OPEN ACCESS
International Scientific and Research Conference on Knowledge-based Technologies in Development and Utilization of Mineral Resources

011002
OPEN ACCESS
Peer review statement

Papers

012001
OPEN ACCESS
Perspectives of comprehensive mineral exploitation based on the principles of public-private partnership

S M Nikitenko, E V Goosen and K S Sablin.....1

012002
OPEN ACCESS
Prospects for application of robotic mechano-hydraulical excavation of gas-bearing coal layers at great depths

V N Fryanov and L D Pavlova.....9

012003
OPEN ACCESS
Study results on geotechnological processes of mining flat thick coal seams with top-coal drawing in face 21-1-5 of "Olzherasskaya" mine

V V Senkus and A Yu Ermakov.....16

012004
OPEN ACCESS
Ventilation of mines developed by the combined method of coal mining

Val V Senkus, A Yu Ermakov and V V Senkus.....26

012005
OPEN ACCESS
Reliability evaluation of the mining flowsheet of seam 21 in "Olzherasskaya-Novaya" mine LLC "UC Yuzhny Kuzbass"

A Yu Ermakov.....29

012006
OPEN ACCESS
A study on directions of significant efficiency increase of rock fracture by tools equipped with super hard inserts from composite

L T Dvornikov, V I Klishin, S M Nikitenko, V A Korneyev and P A Korneyev.....33

012007
OPEN ACCESS
Evaluation of elements loading in the metal structures of powered support units

M S Nikitenko.....43

012008
OPEN ACCESS
Comparison of outburst danger criteria of coal seams for acoustic spectral and instrumental forecast methods

A V Shadrin and Yu A Bireva.....50

012009
OPEN ACCESS
Construction of the approximant of complete diagram for rock deformation

A B Tsvetkov, L D Pavlova and V N Fryanov.....57

012010
OPEN ACCESS
Adjustment of automatic control systems of production facilities at coal processing plants using multivariant physico- mathematical models

V F Evtushenko, L P Myshlyaev, G V Makarov, K A Ivushkin and E V Burkova.....64

012011
OPEN ACCESS
Research into the impact of shafts misalignment of turbocompressor installation on power characteristics of a drive engine

M V Kipervasser and A V Gerasimuk.....73

012012
OPEN ACCESS
Adaptive optimization as a design and management methodology for coal-mining enterprise in uncertain and volatile market environment - the conceptual framework

V V Mikhalchenko and Yu T Rubanik.....80

012013
OPEN ACCESS
Management of local economic and ecological system of coal processing company

T V Kiseleva, V G Mikhailov and V A Karasev.....87

012014
OPEN ACCESS
Construction of a conceptual model of transport system for a coal mining region

Yu D Pristupa, V N Fryanov and L D Pavlova.....94

012015
OPEN ACCESS
On the problem of vehicles assignment in the conditions of open-cast mining

K S Davkaev and M V Lyakhovets.....101

012016
OPEN ACCESS
Modelling of CWS combustion process

I A Rybenko and L A Ermakova.....109

012017
OPEN ACCESS
Rational preparation of waste coal mixture for production of bricks by the method of compression molding

A Yu Stolboushkin, A I Ivanov, M V Temlyantsev and O A Fomina.....118

012018
OPEN ACCESS
A study on sintering of ceramic bricks made from waste coal

A Yu Stolboushkin, A I Ivanov, V A Syromyasov and O A Fomina.....124

012019
OPEN ACCESS
Structure formation of aerated concrete containing waste coal combustion products generated in the thermal vortex power units

A I Ivanov, A Yu Stolboushkin, M V Temlyanstev, V A Syromyasov and O A Fomina.....130

IOP Conf. Series: Earth and Environmental Science **45** (2016) 011001 doi:10.1088/1755-1315/45/1/011001

Preface

Since 1995 Siberian State Industrial University in Novokuznetsk, Russia, within the framework of International Specialized Exhibition of Mining Technologies "Russian Coal and Mining" has been holding an annual scientific conference *Knowledge-based technologies in development and utilization of mineral resources*, which is devoted to the problems of coal production diversification and generalization of the investigation's results, including creation of robotic and automated coal-mining and coal-processing enterprises, based on the use of the advanced technologies in coal and methane mining, complex processing of these products in the coal-mining regions and sales of energy products to consumers in the form of thermal and electric power.

This year the conference was held from June 7 to June 10, 2016. The Organizers of the Conference are: Siberian State Industrial University (Novokuznetsk, Russia), Federal Research Centre of Coal and Coal Chemistry of Siberian Branch of the Russian Academy of Sciences (Kemerovo, Russia), University of Science and Technology Liaoning (Anshan, China), Expo center "Kuzbass Fair" (Novokuznetsk, Russia). The event was actively supported by the Ministry of Energy of the Russian Federation and administration of Kemerovo region and Novokuznetsk city.

The conference is a recognized scientific and practical forum for professional discussions of the problems concerning state and prospects of new technologies and devices development ensuring rational use of mineral resources in the present conditions of socio-economic development of coal-mining regions. The event was attended by scientists and specialists from different institutions, companies and coal mining enterprises from Russia, Germany, Poland, Kazakhstan and the Republic of Belarus. Active participation of PhD students and young scientists is particularly noteworthy.

More than 120 authors and research teams contributed their papers to the conference. The present issue of *IOP Conference Series: Earth and Environmental Science* contains 19 specially selected research papers covering the following topics: scientific bases of new technologies of production, processing and use of mineral resources, design of modern automated control systems of mining production, development and implementation of multi-functional systems of industrial and environmental safety, improvement of forms of economic management of mining enterprises.

The Organizing Committee would like to thank the Russian Foundation for Basic Research for it financial support (project No.16-05-20136 G). We also appreciate contribution of all participants into development of mining science and practice.

Vice-chair of the Organizing Committee
and Editor-in-chief
Prof. Dr. Viktor Fryanov

IOP Conf. Series: Earth and Environmental Science **45** (2016) 011002 doi:10.1088/1755-1315/45/1/011002

Peer review statement

All papers published in this volume of *IOP Conference Series: Earth and Environmental Science* have been peer reviewed through processes administered by the proceedings Editors. Reviews were conducted by expert referees to the professional and scientific standards expected of a proceedings journal published by IOP Publishing.

Perspectives of comprehensive mineral exploitation based on the principles of public-private partnership

S M Nikitenko[1,3], E V Goosen[1,2] and K S Sablin[1,2]

[1] Federal Research Centre of Coal and Coal Chemistry of Siberian Branch of the Russian Academy of Sciences, 10 Leningradsky Ave, Kemerovo, Russia, 650065
[2] Kemerovo State University, 6 Krasnaya Street, Kemerovo, Russia, 654043
[3] Kemerovo Institute (branch) of Plekhanov Russian University of Economics, 39 Kuznetskiy Ave, Kemerovo, Russia, 650992

E-mail: nikitenkosm@ic.sbras.ru

Abstract. The opportunities of the public-private partnership instruments using to shift the paradigm of the natural resources development in the fuel and energy complex of Russia are highlighted in the article. Three main directions of PPP projects development in the subsoil use are pointed out. The role of innovations in the implementation of the most perspective PPP projects in the fuel and energy complex of Russia is shown.

1. Introduction

Fuel and energy complex of Russia plays a significant role in the development of the country. According to the State Balance data Russia is the undisputed world leader in the natural gas reserves; oil and condensate reserves account for 20.1 billion tonnes and ensure the fifth place in the world rankings with a share of 3% of the world reserves. Fuel and energy complex (FEC) of Russia provides about 50% of federal budget revenues, 70% of export earnings and jobs for millions of Russians in mining and related industries. Nowadays it forms the basis and will form the basis in the foreseeable future for the formation of the country's budget [2]. However, the exaggerated role of FEC in Russia has led to the formation of "enclave dual economy", which, according to the well-known American economist G. Stiglitz, is characterized by such features as focus on the extensive extraction of natural resources and the existence of two sectors that are isolated from each other, i.e. highly profitable export-oriented commodity sector is mainly based on the extraction of rents and a lagging underdeveloped manufacturing sector [3]. Enclave economy is weakly receptive to innovative development and hardly moves from the innovation system of "technology push" (the fundamental knowledge generated according to the state efforts) to the innovative system of "market traction" (innovations generated by the efforts of businesses) [4]. It is important to note that a full-fledged internal market has not formed in the enclave economy and created institutions (rules of game) are not effective and conserve prevailing situation [5].

FEC is itself undergoing the negative effects of the enclave economy that consist in the deterioration of the resource base due to the depletion of existing fields, reduction in the size and quality of new geological discoveries, increase in the costs for development of complex and distant provinces. High depreciation of production assets, wide technological gap between Russia and developed countries, high level of dependence on imports of equipment, materials and services as well as excessive dependence on unstable external energy markets are characteristics of modern Russian

FEC. Steady decreasing of demand and energy prices drop in the world market as well as outbreak of geopolitical crisis in 2014 that led to the introduction of sanctions against Russia exacerbated these problems by making it impossible to continue the country development mainly due to the natural energy resources production and sales in the external market. The problem of search for a new model of FEC development is very important. It seems that the solution of this problem is closely connected with the shift from the extensive use of natural resources to the integrated development of natural resources with sustainable socio-economic development of the territories and mutually beneficial partnership of science, authorities and business on the principles of public-private partnership (PPP) [6]. All the factors mentioned above make it actual to search for the perspective directions and forms of PPP directed at the formation of a new paradigm of development of the Russian FEC.

The purpose of this paper is to research into the perspective areas of science, authorities and business partnership on the principles of PPP in the fuel and energy sector adapted to the specific Russian conditions.

2. Discussion of the problem

Despite the fact that the speeches of the country's leadership and the "Energy Strategy of Russia for the period 2035" highlight that with the help of PPP it is planned to create an effective instrument for import substitution and cross-sectoral cooperation, which solve the problem of attracting investments as well as contribute "the formation of scientific, technological and industrial base for the development and production of high-quality power equipment and provide services in the key fuel and energy technology areas for sustainable operation and development" [7], the public-private partnership in the fuel and energy sector has not received wide implementation in Russia. This fact can be explained by three main reasons. Firstly, there is no clear conceptual approach to PPP in Russia. Secondly, national model of PPP in Russia is in infant phase, and legislation, standards and forms of PPP projects are not developed. Thirdly, Russian legislation contains a number of serious limitations of PPP projects implementation in the fuel and energy sector.

Among the studies devoted to the PPP the most influential are the studies of E. Osborne, P. Steven, E. Klijn, G. Teisman, G. Hodge, C. Greve, E.R. Yescombe et al. [8-12]. The following conceptual approaches can be pointed out in the context of foreign and Russian theoretical PPP studies. It is crucial to note that they differ in the understanding of the role and place of the PPP in the modern market economy. According to the first approach PPP is considered in its broadest sense and includes all forms of cooperation between business and authorities, including joint ventures, corporate social responsibility, philanthropy, and even government subsidies and financial and organizational support for business [13]. The broad approach allows the place of PPP in the public sector of economy to be shown, as well as possible directions of joint participation of business and state in the FEC development to be identified. However, this approach hides the features of PPP, its specific mechanisms and forms, benefits and risks. The inability to clearly define the permissible limits and priorities for the use of PPP in the fuel and energy industry is the main drawback of the approach.

The second approach is based on the concept of "New Public Management" (Y. Van Ham, H. Koppenjan, M.B. Gerrard, E.S. Savas, H. Delmone, V.G. Varnavskiy, M.A. Deryabina, N.D. Holodnaya) [13-20]. According to this approach PPP is a way to implement the instruments of project commercial management in traditional branches of the public sector (medicine, public health, protection of public order, public utilities) and to increase efficiency of state property management in the infrastructure sectors, such as transport. PPP is represented as an instrument of state regulation and privatization alternative according to the approach. The main focus of this approach is on the implementation of the approved foreign organizational schemes and ways of financing projects prevailing in developed countries. The main advantage of the approach is the detailed analysis of PPP specifics as a form of interaction between business and authorities, and the main drawback is the implementation of developed countries experience only. Most followers of this approach do not analyze country-specific PPP models and do not consider the possibility of using PPP in fuel and energy complex. One can say with some proviso that this approach is most prevalent in the foreign and Russian literature.

The third approach, being the most perspective for the study of the possibility of using PPP in fuel and energy sphere, considers PPP projects as an instrument for national, international, regional, city and municipal economic and social development, as a way to overcome the economic crisis in some countries and sectors of economy (D.W. Brinkerhoff, J.M. Brinkerhoff, S. Agere, A.E. Kontorovich, S.M. Nikitenko, E.V. Goosen et al.) [6, 20-21]. According to this approach PPP features are clearly defined, i.e. project nature of the interaction (PPP is a long-term project with a clearly defined timeframe); voluntary and mutually beneficial co-operation between partners; formal co-operation between partners on the basis of contracts and agreements with a clear structure of interaction to share risks and benefits; joint participation of businesses and authorities in the financing and management and/or implementation of the project [6]. At the same time the approach admits the possibility of the PPP forms existence in the country or the sector, and that bears the character of a project but do not meet all characteristics of PPP. To distinguish these projects and programs from the "classic" PPP it is supposed to indicate them as quasi-PPP ("almost" PPP) [22]. This approach can exactly identify the most perspective areas and features of PPP development in the fuel and energy sector in Russia.

Let us consider the Russian experience of the PPP partnership development. PPP projects market began to develop in Russia after the formal adoption of the Federal Law from 21.07.2005 No. 115-FZ "On Concession Agreements" (hereinafter the Law on Concessions). There was a rapid increase in the number of projects and investments for the period 2005-2015. According to the official portal "Unified information system of public-private partnership in the Russian Federation" (hereinafter PPP Information portal) by the middle of 2016 decisions had been made to implement 1339 projects, and 873 of them are being implemented at the moment. The number of projects has increased almost 10 times compared to 2014. Table 1 indicates the dynamics of the registered projects in Russia. It should be noted that in the Siberian and Far Eastern Federal Districts, where the main resource-extracting companies in Russia are located, the growth in the quantity of projects was the highest [23]:

Table 1. Dynamics of quantity of PPP projects in the Russian Federation according to the PPP Information portal (2013-2015).

District	Number of PPP projects by year			Growth from 2014 to 2015 (times)
	2013	2014	2015	
Central Federal District	59	21	292	**14**
Northwestern Federal District	46	23	114	5
Volga Federal District	36	34	342	10
Southern Federal District	14	10	61	6
North Caucasian Federal District	9	4	28	7
Ural Federal District	25	11	49	4.5
Siberian Federal District	103	24	256	**11**
Far Eastern Federal District	19	4	143	**36**
Russian Federation (total)	311	131	1285	**9.8**

Another feature of the official Russian market of PPP projects is that the projects of the fuel and energy sector are almost not represented. Only five projects out of 1285 can be attributed to the sphere of subsoil use in 2015 after the Federal Law of 13.07.2015 N 224-FZ "On public-private partnership, municipal-private partnership in the Russian Federation and the introduction of changes into the Russian Federation legislative acts" (hereinafter the PPP Act) was introduced in 2016. PPP projects in the sphere of subsoil use were excluded from the database as not corresponding to the national legislation after adoption of the amendments to the Law on Concessions.

According to the quality of the PPP projects and the volume of attracted investments the Russian PPP projects also do not entirely correspond to the main features of the classic PPP projects. Most PPP projects are short-term in the contrast to the country-leaders such as the USA, Great Britain and Canada. Almost 30% of PPP projects that are registered in the PPP Information Portal have a 3-year implementation period or less, while in most countries implementation duration of PPP projects is 10 or more years. Russian projects have differences in the mechanism of financing. An unusually high proportion of investment made by the state characterizes Russian PPP projects. If in the developed countries this share usually is less than 25%, then Russian successful PPP projects get more than 50% of budgetary investments.

This is the reason why the total share of private investment in PPP projects in Russia is not only much lower than in the developed countries but also lower than in the BRICS countries, i.e. in Brazil and India. Total private investment in PPP projects in Brazil and India represents 18.89% and 9.47% of GDP respectively, then in Russia the share of the total private investment is less than 1% as shown in Figure 1 [23]. Many Russian PPP projects do not differ from ordinary projects implemented by means of budgetary investments on free and irrevocable basis, while the classic PPP projects involve an investment mechanism for raising funds through issuance of bonds with the use of private financial institutions and funds. It is not surprising that many Russian researchers study PPP projects only as a way to attract budget investments [23]:

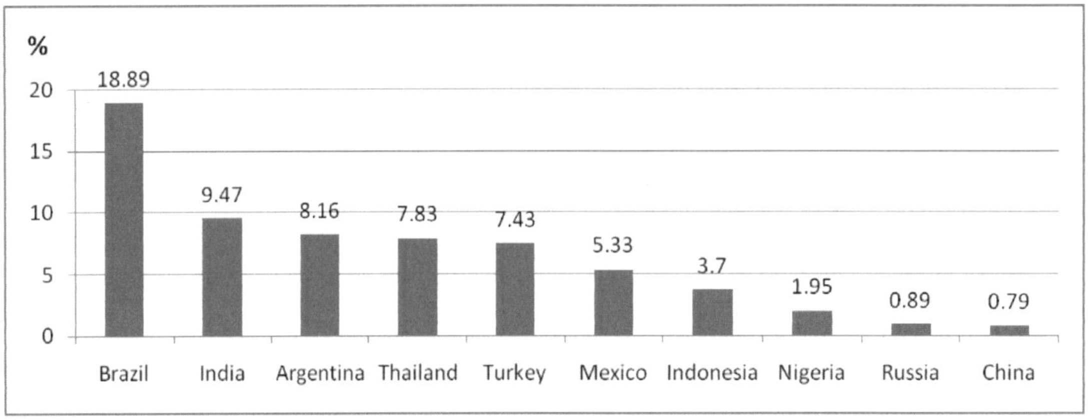

Figure 1. Share of private investment in PPP projects to nominal GDP (%).

A major factor that limits the development of PPP projects in Russia in general and, particular, in the subsoil use is the presence of a large number of quasi-PPP projects in the actual practice. Infrastructure projects that are implemented on the basis of the State order and agreements on social and economic cooperation between regional authorities and businesses as well as federal and regional programs considered as PPP projects according to some regional laws on PPP. Some Russian regions consider ordinary business support programs as PPP. Their formation and operation are regulated by the general rules of civil law, that is not adapted for this, or even these programs are outside of the legal field at all.

Special place among quasi-PPP occupy agreements between authorities and businesses about socio-economic development as well as large mining-extracting companies programs of regions and territories development that are widely used in natural resources extracting regions. Table 2 from [23] contains a complete list of such quasi-PPP forms projects that are included in the PPP Information portal database. Column "quantity" provides data about the number of specific forms of the projects included in the database. However, these figures do not reflect the real picture because most of these forms of partnership between business and authorities is not legally regulated, which prevents from recording and tracking them officially. Earlier studies fulfilled by the authors have shown that the number of quasi-PPP projects is a hundred times greater than the number of classical PPP projects realized in Russia [24].

Table 2. Forms and quantity of quasi-PPP projects included in PPP-info database (beginning of 2016).

Forms	Quantity
Investment agreements	16
Long-term investment contracts with an investment component	1
Credit agreements with Vnesheconombank	1
The contract of uncompensated use of property	2
Agreements on social and economic cooperation	7
As a part of the non-commercial organizations law	1
Production sharing agreements	5
Overall quasi-PPP projects (share of the total number of projects)	28 (2%)

Despite the fact that most of the Russian projects are concession agreements in the infrastructure and social sphere, there is a positive experience in the implementation of PPP projects in industries engaged in production, processing and transportation of natural resources. Examples according to [23, 25]: building transport infrastructure for the development of mineral resources of the south-east of the Chita region, construction of the railway line Kyzyl – Kuragino in conjunction with the development of mineral resources base of the Republic of Tuva, a program of integrated development of the Lower Angara region (see Table 3).

The share of such projects is small. These projects are implemented within the large programs of sectors and regions development and are financed by the means of the Investment Fund of the Russian Federation and do not officially belong to the PPP projects. Two Russian official PPP projects database – Information Portal of the PPP and the official website of the Ministry of Economic Development, that register projects financed with the participation of the Russian Federation Investment Fund, provide information only on 12 PPP projects. Projects related to subsoil use amount less than 1% of all PPP projects registered on both websites. At the same time only 6 projects can be directly attributed to the fuel and energy sector (Table 3).

Table 3. Russian PPP projects in the fuel and energy sector.

No.	Project name	Terms of realization	Industry classification	Project status
1	Elaboration of project documentation for the investment project "Integrated Development of South Yakutia"	2008-2013	coal mining, enrichment, energetics	federal
2	Building of transport infrastructure for the development of mineral resources of the south-east of the Trans-Baikal Territory	2007-2016	enrichment, transportation	federal
3	Complex of refineries and petrochemical plants in Nizhnekamsk	2006-2012	petroleum refining	federal
4	Integrated development of the Lower Angara region	2006-2015	transportation, energetics, non-ferrous metallurgy	regional
5	Creation of a complex for processing Northern Caspian gas into ethylene, polyethylene and polypropylene (I step)	2011-2015	gas processing, energy and transport infrastructure	regional
6	Construction of the railway line Kyzyl-Kuragino in conjunction with the development of mineral resources base of the Republic of Tyva	2008-2016	transportation	regional

A small number of projects in fuel and energy sector of Russia are largely due to legal limitations. For instance, Part 1 of Art. 3 (PPP Law) established a closed list of objects that can be a subject of the PPP agreement. PPP Law prohibits expressly the conclusion of PPP agreements in respect of natural resources. Licensed character of subsoil use also limits the ability to create PPP projects in the fuel and energy sector and in the framework of the Law on Concessions. Another legal barrier to the development of PPP in the fuel and energy sector is the definition of a private partner in the PPP projects that is given in Part 3 of Art. 3 of the PPP Law, according to which the state and municipal unitary enterprises, state-owned companies as well as non-profit organizations and "offshoots" of state companies can not be private partners in PPP projects. This limitation automatically makes it impossible to participate in PPP projects for most big oil companies, which significant percentage of shares belong to the state.

Organizational structure of the companies from the fuel and energy complex limits seriously the possibilities of PPP development. Most of the regional fuel and energy companies are not independent, they are part of the large vertically integrated groups that were registered and have profit centers in Moscow and St. Petersburg. Parent company accumulates most of the revenues of its regional companies as well as most of the taxes through transfer pricing system in profit centers that limits severely the ability of the implementation of regional PPP projects. It is fair to say that the situation with a relatively small share of PPP projects in the natural resources development is not a specific feature of Russia but is a typical feature for the majority of the countries.

Despite the highlighted problems, the PPP projects in the subsoil use as well as fuel and energy complex can be actively generated in three main directions. The first direction is the creation of specialized production of mineral-raw material orientation on the basis of concession agreements and production sharing agreements. As an example of such PPP in Russia can serve the development projects of Elgiyskiy coal deposit (Republic of Sakha, Neryungri district), oil and gas fields of Evenkia (Yurubcheno-Tokhomskoye, Kuyumbinskoye, Nizhneangarsk group, Sobinsko-Teterinskaya group).

Projects of hydrocarbons development in the form of production sharing agreement with the participation of foreign investments can be included in the PPP in the sphere of fuel and energy sector but with some proviso. These projects are: Sakhalin-2, which includes the Piltun-Astohskoe oil and gas condensate field and Lunskoye gas condensate field; Sakhalin-1, including Chaivinskoe, Arkutun-Daginskoe and Odoptinskoe oil and gas fields; "Khariyaga" project; Samotlor oil and gas condensate field development project. It is important to note that this type of PPP projects does not seriously change the paradigm of subsoil use. These projects serve to overcome the crisis as well as an instrument to support the current state of the industry [26].

The second direction is the creation of industrial and social infrastructure in the resource-extraction regions, forming conditions of rational subsoil use, deep processing and enrichment of extracted resources. As a striking example of this PPP project in Russia can serve the construction of the complex of refineries and petrochemical plants in Nizhnekamsk. The initiators of the construction were the Government of Tatarstan and the company PJSC "Tatneft", which became the main private partner, investor and project coordinator.

The aim of the project was the construction of processing enterprises to refine Tatarstan oil near the site of its production; substitution of oil export with realization of high-quality oil products on the domestic and foreign markets that consistent with the strategic objective of Russia; improvement of the environment through the production of environmentally friendly fuels and compliance with stringent emission requirements for the design of the Complex installations; application of the advanced world technologies. Integration of refineries and petrochemical plants into a single production facility will give impetus to the development of intra- and inter-regional integration of the region's businesses.

At the first stage of the PPP project realization over 3,000 new jobs were created and apartments for workers were built in the framework of the project as well as kindergarten and a sanatorium complex [22, 26]. It is important to indicate distribution of "responsibilities" between partners. PJSC

"Tatneft" as a private partner was engaged in construction of production facilities using its own funds, whereas state upgraded external infrastructure using means of State Investment Fund: access rail tracks, oil pipeline with pumping station and pipeline for finished products. Approximately 16.5 billion rubles of budget investments from the Investment Fund were allocated for this purpose. At the same time the total cost of the project amounted to over 200 billion rubles. Construction of the plant was started in 2006 and now is in the final stage [26]. PPP projects of the second direction have a significant social component. They provide a large-scale modernization of the transport, energy and social infrastructure and aim at the socio-economic development of the region (increasing employment, standard of living). That is why the PPP projects are much closer to the paradigm shift of subsoil use, but do not change it fundamentally.

The third direction is the integrated development of natural resources. This direction is associated with the projects of new industries formation that focus on creating innovative development centers, developing markets and clusters. Such projects are still very few in the world. PPP projects of the third direction are diversified innovative projects for the fuel and energy industry, including development of innovation centers, technology platforms, etc. The main difference of these projects is due to their focus on the search and selection of basic research oriented ideas, conducting research based on scientific studies, research and development of innovative business projects.

Three technological platforms for the extraction of natural resources and oil and gas refinery approved by the Government Commission on High Technology and Innovation in 2011 can serve as examples of PPP projects of the third direction in fuel and energy complex. Twenty second technological platform is a platform of solid minerals, focused on deep processing and renewing domestic industries with high technologies. The 23rd – technology of hydrocarbon mining and utilization, the result of which should be the development and implementation of new technologies for the production, preparation, processing and transportation, well drilling, etc.; the 24th – deep processing of hydrocarbon resources, which implies the creation of conditions for technological modernization and increasing competitiveness of oil and gas refining, petrochemical industry and organic synthesis procedures using foresight [26]. The main objective of technological platforms is to select the ideas of fundamental research, performance of scientific research on their basis and development of innovative business strategy. As a form of PPP development, it not only creates a fundamentally new technology, but also involves business and research organizations into the process of formation and development of new markets on the basis of close mutual cooperation.

3. Conclusions

The performed analysis shows that the PPP projects in the fuel and energy sector in Russia are developing very slowly. However, PPP projects that focus on innovative model of integrated development of natural resources have potentially great perspectives. These PPP projects can solve fundamentally the problem of resource-extracting regions due to the transition to a new paradigm of subsoil use based on the idea of the integrated use of natural resources and sustainable socio-economic development of territories on the basis of new technologies introduction, formation and development of the internal market.

4. Acknowledgements

The research was supported by the Russian Scientific Fund (research project No. 16-18-10182 "Formation of the organizational-economic mechanisms of integrated development of mineral resources based on the partnership between science, business and authorities in the resource type regions").

References

[1] *State Report on the Status and Use of the Russian Mineral Resources in 2014* 2015 (Moscow)

[2] Stiglitz J E 2002 Towards a New Paradigm for Development: Strategies, Policies and Processes (Columbia University: Academic Commons) http:// hdl.handle.net/10022/AC:P:19637

[3] Rutskiy V N and Pyzhev I S 2015 *J. of Institutional Studies* **7** (4) 128–137
[4] Levin S N, Sablin K S, Kagan E S 2015 *J. of Institutional Studies* **7** (3) 92–101
[5] *Status and Prospects of Development of Public-private Partnership Projects in the Context of Integrated Development of Mineral Resources* 2015 (Moscow-Kemerovo) ed A E Kontorovich, S M Nikitenko and E V Goosen p 331
[6] *Energy Strategy of Russia for the Period 2035 (Main Provisions)* 2015 (Moscow)
[7] Delmon J 2009 *Private Sector Investment in Infrastructure: Project Finance, PPP. Projects and Risk* (The World Bank and Kluwer Law International)
[8] Osborne S and Steven P 2000 *Managing Public-Private Partnerships for Public Services: An International Perspective* (London: Routledge) p 268
[9] Klijn E H and Teisman G R 2000 *Governing Public-Private Partnerships: Analyzing and Managing the Processes and Institutional Characteristics of Public-Private Partnerships* (Public-Private Partnerships: Theory and Practice in International Perspective) ed S P Osborne (London: Routledge) pp 84–102
[10] Hodge G A et al 2010 *International Handbook on Public-Private Partnerships* (Cheltenham, UK: Edward Elgar) p 631
[11] Yescombe E R 2015 *Public-private Partnership: The Basic Principles of Finance* (Moscow: Alpina Publisher, in Russian) p 457
[12] Gerrard M B 2001 *Finance & Development* **38** (3) 27
[13] Ham H van and Koppenjan J 2001 *Public Management Review* **3** (4) 598
[14] Savas E S 2000 *Privatization and Public Private Partnerships* (N. Y.: Chatham House Publishers, Oxford University Press) p 368
[15] Delmon D 2010 *Public-private Partnership in Infrastructure* (Astana: Apelsin) p 250
[16] Varnavskiy V G 2011 *World Economy and International Relations* **9** 41–50
[17] Deryabina M 2008 *Voprosy Ekonomiki* **8** 61–77
[18] Holodnaya N D 2009 *Problems of State and Municipal Management* **2** 42–56
[19] Brinkerhoff D W and Brinkerhoff J M 2004 *International Review of Administrative Sciences* **70** (2) 253–270
[20] Agere S 2000 *Promoting Good Governance: Principles, Practices and Perspectives* (London: Commonwealth Secretariat) p 144
[21] Goosen E V, Nikitenko S M and Pahomova E O 2015 *EKO* **2** 163–175
[22] *Unified Information System of PPP in the Russian Federation* http://www.pppi.ru
[23] Goosen E V 2013 *Scientific Papers of Donetsk National Technical University* **1** (43) 121–134
[24] *Reference Materials on the Projects Implemented with the Involvement of the Russian Federation Investment Fund* 2015 (M.: Ministry of Economic development of the Russian Federation)
[25] Polyanskaya I G 2009 *Business, Management and Law* **1** (18) 46–50

IOP Conf. Series: Earth and Environmental Science **45** (2016) 012002 doi:10.1088/1755-1315/45/1/012002

Prospects for application of robotic mechano-hydraulical excavation of gas-bearing coal layers at great depths

V N Fryanov and L D Pavlova

Siberian State Industrial University, 42 Kirova Street, Novkuznetsk, Russia, 654007

E-mail: zzz338@rdtc.ru

Abstract. In the paper the scientific basis for robotic technology of excavation prone to gas-dynamic phenomena of a coal layer with high gas-bearing capacity at great depths is developed. The constructive scheme of automated remote-controlled mining robot, that fractures coal by high-pressure hydraulic jets in the gas polluted environment, is proposed.

1. Introduction

The increased competition on the international market of energy resources due to falling prices for oil and natural gas has led to a reduction in production volumes and consumption of coal. Therefore, to compensate for the declining revenues of the oil and gas industry in the world markets the consumers of oil and gas are searched for on the domestic markets. To create an appropriate niche in the energy system of Russia the oil and gas business in the country proposes to introduce a "carbon tax" with a starting rate of $ 15 per ton of CO_2 emissions. One of the directions for formation of the specified niche is a reduction in the proportion of consumption of coal and production of thermal and electric energy using alternative energy sources.

Supporters of creating "carbon-free zones" by reducing coal production and its replacement by alternative energy sources have strong argument:

- the resolution of the climate summit 2015 in Paris on compulsory reduction of emissions of harmful gases, including in Russia in the next 15 years by 30% compared to emissions in 1990 to reduce the rate of warming of the planet;
- high accident rate at coal mines of Russia – for the last 12 years there have been 10 major accidents and the death toll reached 391.

These factors have led to a reduction in the share of coal in the country's fuel and energy balance to 11-12%. However, the share of coal is significantly higher in developed countries: USA – 18%, UK – 21%, Japan – 30%, which provides greater security to coal-fired power industry to nuclear and hydropower.

In the current competition the first step of the Russian Government on the path of coal mining reduction is to liquidate vary gassy mines. The main cause for conservation and liquidation of mines is an increased risk of gas-dynamic phenomena and endogenous fires in case of spontaneous combustion of coal layers. The nature of these events is poorly understood especially in areas of high seismic activity [1, 2].

One of the variants for increasing industrial safety of coal mines and reducing the risk of incidents dangerous for the staff is to reduce the number of employees by replacing them with machines-robots in hazardous areas. In this regard, it is relevant to develop scientific basis of underground geotechnology providing an increased efficiency of underground coal mining and industrial safety through the use of robotic excavation of gas-bearing coal layers at greater depths.

The purpose of the paper is to scientifically justify the technological system of robotic mines, including remote coal excavation by hydraulic method, hydraulic transportation of the mined rock, underground dewatering of coal slurry.

2. Description of the technological and technical solutions for the design of robotic mine

The authors developed and proposed the main provisions of robotic technology for underground coal extraction which can be used in the design of a new type of mines [3-4, 13-14]. Justification of the structure and elements of robotic technology of underground coal mining was carried out through the realization of the basic principle: the absence of a person in hazardous areas and the implementation of the functions of the operator in a comfortable environment, remote control of the robot during the excavation, coal excavation, the rock mass transportation to the main workings.

The methods and schemes of opening and preparation of a mine field of robotic mine are accepted conventional recommended by normative and methodological documents, e.g., [1, 7]. The need to develop new methods and schemes of opening and preparation of mine fields within the technological system of robotic mine in the first stage of its creation does not arise, since the safety of work in the performance of these processes is provided by the mines with traditional technology within the level of acceptable risk. As the largest risk to personnel present mining operations within the extraction area, the proposed technological solutions in the robotic structure of the mine must ensure the safety of mining operations during preparatory workings, stope excavation of coal and transportation of rock mass.

The following technical and technological solutions are incorporated in the proposed robotic technology mine system [1, 7]:

1) Exclusion of staff presence in clearing and preparatory coal faces by means of a remote control of robotic coal mining from operator stations located outside the areas hazardous for humans.
2) Performance of basic coal extraction operations in the methane-air environment with an active dust control and water supply on the mechanism of coal fracture.
3) Hydrotransportation of ROM with an underground dehydration cycle of rock mass, clarification of water and waste management in the goaf.
4) Performance of development workings on the pre-drilled wells of large diameter with ventilation of shaft faces due to the mine depression, seam degassing and unloading of the array of rock from pressure to prevent rock bursts and sudden coal and gas outbursts.

The basis for the implementation of technical solutions in technological system of a robotic mine is a remote-controlled self-propelled robot. The main robot functions are dredging coal seam according to a given program and monitoring of the following stope elements: shape and size of the edge part of a seam, stability of the roof rocks, the parameters of mine atmosphere, ROM transportation from the stope. All information on the status of these stope parameters is transmitted to the operator to adjust the program.

The possibility for creation of such robot is provided by scientific and technological developments in the field of robotics, development and implementation of remote sensing of geomassif condition and process monitoring. As a counterpart in the design of the robot control system it is proposed to use technical devices and methods proposed in [8, 9].

The main technological element of an excavation robot is a method of coal fracture and transportation. There are many variations of designs of mechanical, hydraulic, pneumatic, vibrating coal fracture [10, 13]. Given the need for a comprehensive solution to the problem of fracture, excavation, loading, ROM transportation, one energy source is adopted – high-pressure water, which will ensure the coal fracture and rock mass jetting, dust control, reduction of the probability of explosion of methane mixture, transportation of coal-water slurry to the underground dewatering chamber [10, 11].

In the course of development of hydraulic underground coal mining a number of methods was proposed and tested in practice, which are fracture of coal by water jets, including: thin jets of water with pressure up to 60 MPa, middle-pressure jets with water pressure up to 12 MPa and a flow rate 300 m^3/h.

The latter option was most widely used in hydraulic mines, however, because of the high energy capacity it was not cost-effective compared to the mechanical method of destruction, and was replaced by a combined (mechano-hydraulical) method with a mechanical coal fracture, ROM jetting on the soil of working and the formation of a coal-water flow.

According to the research it was found that the performance of a hydraulic rock fracture greatly depends on the distance from the hydrojet nozzle to the face. In [12] it was proved that as the distance from the nozzle to the surface of the coal seam increases in the range of 30 ÷ 300 mm, the hydraulic fracture productivity decreases almost 50 times. Therefore, continuous and efficient performance of the excavation robot can be achieved by the approximation of its fracturing mechanism to the changing face line at the optimum distance.

The distance from the face to the nozzle is 30 – 50 mm [13], and the possibility of its realization can be achieved by automated robot control. Efficacy of fracture is increased during swings of a cutting mechanism equipped with several nozzles with overlapping jets of water. Another important parameter of the fracturing mechanism of the robot is water pressure, with its increase in the range 10 ÷ 40 MPa the coal fracture rate increases 5 times [13].

Thus, the results of theoretical and experimental research confirm the possibility of a robotic excavation performance of one unit within 50 ÷ 80 tonne/h with program control of cutter motion.

According to the technological scheme of a robotic mine it is necessary to ensure the work of excavation robot in three modes: execution of a mine working with section 1.0 ÷ 1.5 m^2 in the direct motion, widening of a mine working up to 6 m in the reverse motion, liquidation of roof rock collapses, rock mass congestion near the robot and other emergencies.

Selection of the required minimal cross-sections of the mine workings was made in accordance with the production experience of application of hydraulic method [10] and result of research of the influence of underground workings form and sizes on rocks stability [5]. Given that the use of robotic technology is possible in difficult geological conditions unfavorable to traditional technology, the round shape of the mine working with small cross-section is adopted. This will allow the mine workings to be conducted without support and the probability of failure of roof rocks to be reduced. In case of local roof rocks fracture in the area of mine working, including behind the excavation unit, the design and the automatic control system provides the possibility for fracture of coal and oversized pieces of rock.

The design of an excavation robot is shown in Figure 1. It consists of the following main components:

1 – body for the assembly and configuration of main components, high-pressure water supply and channels for exchanging information between the operator and the robot;

2 – spacer-chassis for the robot movement, that provides its stability during work and movement in the circular mine working and at the junction with the chamber (Figure 2);

3 – executive mechanism of coal fracture in the form of jet forming box, frame and hoses for water supply and control actions to move the nozzles;

4 – executive mechanisms for the fracture of oversized pieces and accumulations of coal and rock;

5 – controlled overlapping in order to prevent the roof rock fracture;

6 – hoses and connections for transmitting and receiving information;

7 – flexible tube for transmitting power;

8 – system of remote and automatic control.

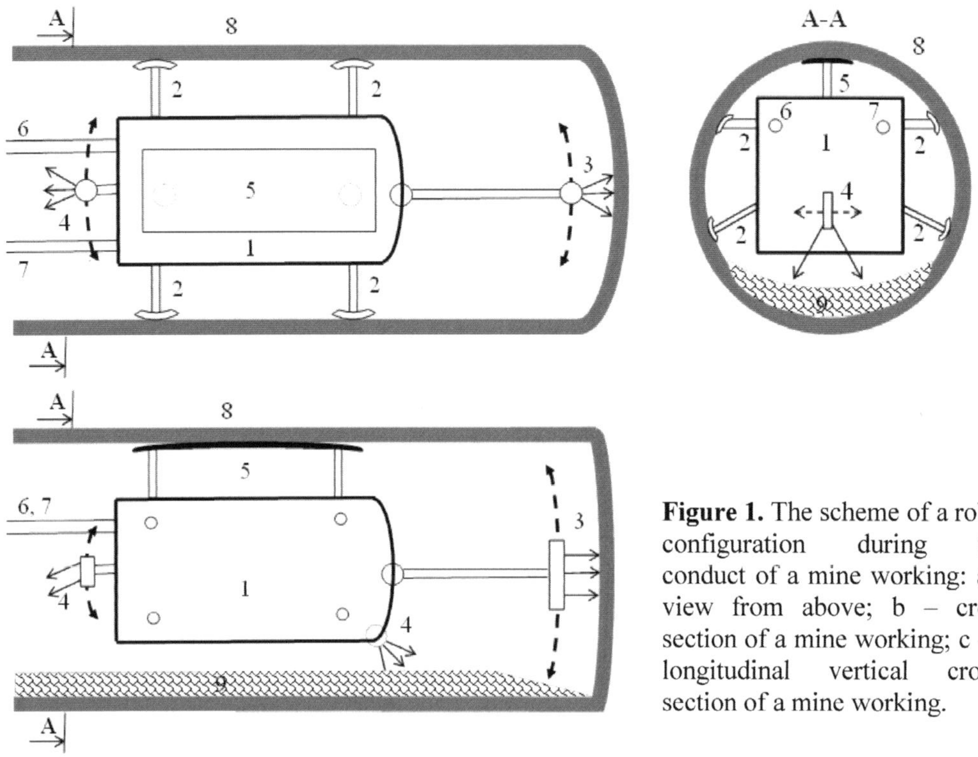

Figure 1. The scheme of a robot configuration during the conduct of a mine working: a – view from above; b – cross section of a mine working; c – a longitudinal vertical cross-section of a mine working.

Besides the offered nodes in the construction process other options of excavation robot configuration are possible. For example, it is recommend to use compact tunneling machines with the developed [6] original constructions of vehicles with the supports into the roof, ground or main workings sides of various shapes, sections and directions using a spacer-stepping-sliding mechanism.

As a basis for the creation of an automated control system solutions of domestic and foreign companies can be used that produce automated remote control mining equipment.

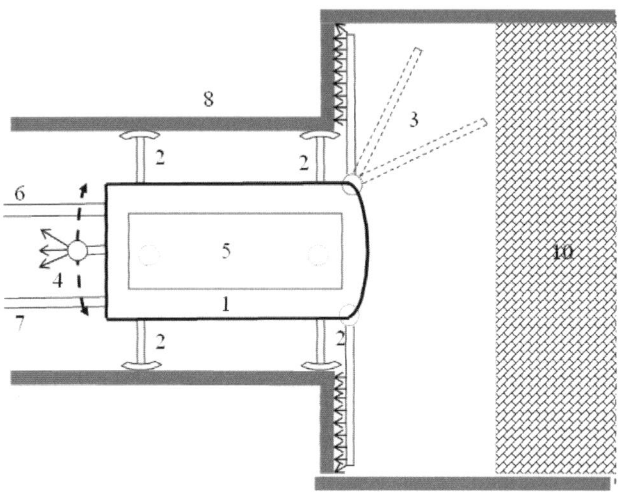

Figure 2. Coal extraction scheme in the chamber by a robotic unit.

One of the options of opening, preparation and mining of the coal layer area on robotic technology is shown in Figure 3.

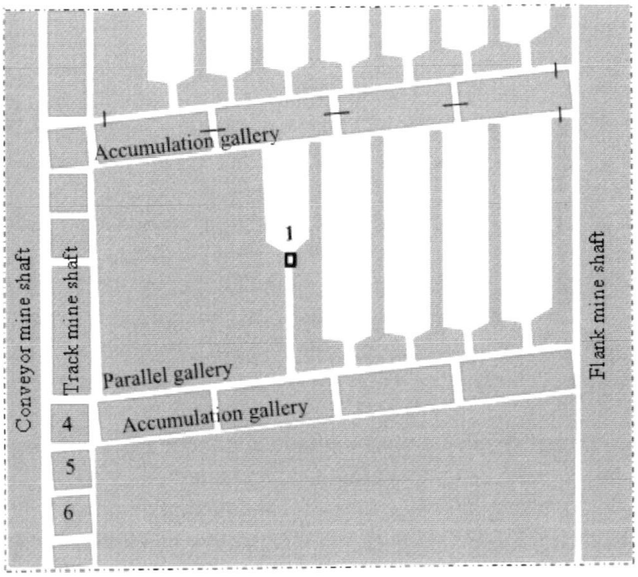

Figure 3. Variant of the development system with automated coal extraction.

The coal layer with the angle of incidence of 35 ° is opened by inclined central and flanking mine shafts, which are executed in a traditional way. The scheme coal layer development is panel. The panel is prepared by parallel and accumulating galleries. Galleries are connected by cross-cuts for airing of the development faces, exit of people in emergency situations and slurry transportation. For gravity hydraulic transportation of coal slurry the galleries are conducted with a slope 0.05-0.07. The panel length is limited by the technical characteristics of the extraction robot and is 150-250 m.

3. Results of designing of a technological system for robotic coal layers excavation

To implement robotic underground coal mining technology the variant of system development is offered, which includes both standard technological solutions and original, related to the characteristics of the processes of excavation sites preparation, fracture and transport of coal. Technological scheme of a production area at a robotic mine is developed on the basis of a principle of organization of a complex production and organizational system. For automated control of robotic systems in complex geological conditions it is necessary to create a model of the control object. For this the following steps of a model creation should be performed:

- a comprehensive description of the system;
- study of the problem situation with the list of elements and situations;
- development of the functional schemes of basic processes interaction in the production areas;
- identification of the field for model application;
- formation of a conceptual model;
- algorithmic description of processes and operations;
- preparation of an information description of the model;
- construction of a formal model.

The comprehensive description of the robotic excavation site includes the following main assemblies of subsystems and processes:

- preparation of coal reserves for excavation, including the methods and technical means for workings development, transportation of mined rock, airing of preparatory faces, degassing of

IOP Conf. Series: Earth and Environmental Science **45** (2016) 012002 doi:10.1088/1755-1315/45/1/012002

coal layers in the area of preparatory workings, discharge from rock high voltage, industrial safety;

- development of methods for substantiation of clearing faces parameters, methods and means of workings excavation, roof control, robotic coal mining with automation and remote control of technical devices;
- methods and means of rock mass transportation;
- methods and means of servicing the technical equipment and staff life support during repair works;
- energy supply of clearing and preparatory faces;
- communications, signaling, visualization of processes and operations of automated remote control.

Near conjunction of central inclined shafts and accumulation gallery chambers of pumping units are built, the rock mass dehydration and clarification of technological water (4-6 in Figure 2). From the central shafts the parallel and accumulating galleries are located. Airing of these galleries is carried out due to mine depression.

To excavate coal by the robot the workings of circular cross-section are conducted. Since the presence of people is not required in the working face and there are no power sources, it is assumed to use the excavation robot in the dangerous gaseous environment. The operators location on the parallel and accumulating galleries is ventilated due to the mine depression.

After the chamber reaches the upper gallery the executive mechanism of the robot by the operator command turns in accordance with the scheme presented in Figure 2, and coal mining is conducted in a reverse motion. Transportation of the rock mass is carried by water to the dehydration chamber and by conveyor to the Earth's surface.

Since coal excavation is carried out near the edge of the erodible layer, the flow of the coal slurry will be formed near the conjunction of the working and camera. Under these conditions, the collapsed rock (10 in Figure 2) will not float to the robot, which will reduce the ash content of the rock mass. By operator command the axis position of the fracturing mechanism must be changed depending on the angle of dip, flow rate and capacity of the cross-section under the the robot body.

For the parallel development workings in particularly difficult conditions (gas-bearing layers prone to gas-dynamic phenomena) it is offered to use a robotic excavation complex according to the scheme shown in Figure 4.

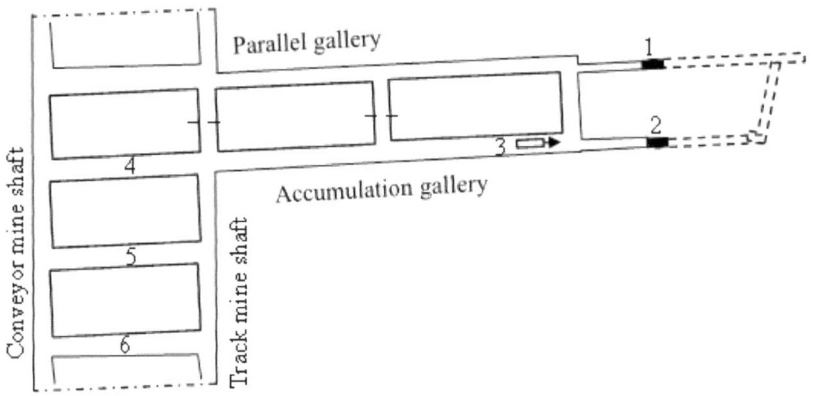

Figure 4. Scheme of the parallel galleries drifting with unloading and degassing of rocks by wells with diameter more than 1 m.

Wells of large diameter are located ahead of the development faces in the coal layer, which help to degas and unload rocks. Simultaneously with the airing of wells due to the mine depression the heading machine (3 in Figure 4) widens the mine working to a full cross-section with a hydraulic transportation of rock mass. This method decreases dustiness in the development face, coal dust and methane-air mixture are removed by advanced well, restrictions on the gas factor are eliminated. This leads to an increase in the speed of developments and provision of comfortable working conditions.

4. Conclusion

On the basis of the developed technological and technical solutions for design of robotic mines and their discussion the following conclusions and recommendations for the practical use of the results are proved.

1) A promising direction for the increase of industrial safety at coal mines and reduction of the risk of accidents is the development and construction of robotic coal mines.

2) The main differences of the robotic coal mine compared to the traditional one are:

- exclusion of staff presence in the stopes and development faces by means of a remote control of robotic coal mining from operator stations located outside the areas hazardous for humans;
- performance of major coal mining operations in the methane-containing atmosphere with active dust control;
- drifting of mine workings with advanced wells of large diameter and airing of development faces due to the mine depression.

3) The mandatory elements of construction of an excavation robot are the following key components: body, chassis, the executive mechanism of coal fracture, jet forming box, controlled overlap, hoses and connections for transmitting and receiving information, flexible stands for the electricity transmission system and automated remote control.

References

[1] Ruban A D 2010 *Preparation and Development of Coal Layers with High Gas-bearing Capacity* (M.: Book Mining) p 500
[2] Zubkov V V 2013 *Mine Surveying and Subsurface Use* **6** (68) 22–24
[3] Fryanov V N and Pavlova L D 2011 *Proc. of Conf. on High Technologies of Development and Use of Mineral Resources* (Novokuznetsk: SibSIU) pp 39–50
[4] Pavlova L D, Fryanov V N and Kornev E S 2011 *Proc. of the All-Russian Conf. on Geodynamics and Stressed State of Bowels of the Earth* (Novosibirsk: Institute of Mining of the SB RAS) Vol 2 pp 163–169
[5] Fryanov V N and Pavlova L D 2016 *Proc. of the IV Russian Scientific Conf. on Simulation and High Information Technology in Technical and Socio-economic Systems* (Novokuznetsk: SibSIU) pp 21–27
[6] Shernix N G 2012 *Coal* **11** 48–51
[7] Bratchenko M I et al 1985 *Methods of Opening, Preparation and Mining Field Development System* (M.: Subsoil) p 494
[8] Panzhin A A 2013 *Mine Surveying and Subsurface Use* **6** (68) 32–35
[9] Besedina A N, Kabychenko N V and Kocharian G G 2013 *Physical and Technical Problems of Mining* **5** 20–36
[10] Muchnik V S, Holland E B and Marcus M N 1986 *Underground Hydraulic Coal Mining* (M.: Subsoil) p 223
[11] Atrushkevich A A, Kaido I I and Fomichev S G 1999 *Coal* **10** 63–64
[12] Nikonov G P, Kuzmin I A, Ischuk I G and Goldin Y A 1973 *Scientific Bases of Hydraulic Coal Destruction* (M.: Nauka) p 147
[13] Pavlova L D and Fryanov V N 2012 *Proc. of Int. Scientific Conf. on High Technologies of Development and Use of Mineral Resources* (Novokuznetsk: SibSIU) pp 98–103
[14] Fryanov V N and Pavlova L D 2013 *Proc. of Int. Scientific Conf. on High Technologies of Development and Use of Mineral Resources* (Novokuznetsk: SibSIU) pp 5–11

Study results on geotechnological processes of mining flat thick coal seams with top-coal drawing in face 21-1-5 of "Olzherasskaya" mine

V V Senkus and A Yu Ermakov

LLC "Siberian Scientific Research Institute of Coal Processing", 1 Gornaya Street, Prokopyevsk, 653000, Russia

E-mail: senkusvv@suek.ru

Abstract. In the paper the modes of hydraulic props of powered support are considered; the regularities of their loading are established that allows the powered support for thick flat seams mining with top-coal drawing to be chosen.

1. Introduction

The standard longwall mining technology requires the extraction of the lower seam layer using a mechanized complex and drawing the top-coal (of average thickness 3.7 m) on the goaf conveyer located under the powered support. The length of top-coal draw is 7-8 meters (5 units).

The technological process of coal extraction with top-coal drawing in the longwall face includes the following operations: angle shearing; extraction of the coal "chip"; moving the support units; moving the face conveyor; top-coal drawing; moving the goaf conveyor.

Top-coal is drawn after each pass of the shearer, right after the support units shifting. The process of top-coal drawing onto the goaf conveyor is as follows:

- by the reduction of the telescopic cylinder gate of the tail barrier the top-coal is drawn onto the conveyor;
- by lifting and lowering the tail barrier the additional softening and crushing of the top-coal is achieved, while the coal drawing and loading onto the conveyor take place;
- extension of the gate allows large lumps of coal to be crushed.

If ROM contains up to 30% of rock, the gate is closed and the coal draw is stopped. Large lumps of rock are pushed from the conveyor by the tail barrier. In case of partial coal drawing the unit is moved down two or three times at the height of 200-300 mm and to its maximum setting load for crushing the top coal.

The coal bench is not released at a distance of 5m at the top and bottom parts of the longwall face.

After finishing the top-coal draw operation the goaf conveyor is shifted. The shifting length at one time should not exceed 22.5 m (15 units) and should be less than 7.5 m (5 units).

2. The research results and their analysis

The presence of the ruptured zone and rock-mass weakening in the area of installation chamber caused the smooth roof rock collapse and timely roof coal caving during dismantling the complex from the installation chamber.

IOP Conf. Series: Earth and Environmental Science **45** (2016) 012003 doi:10.1088/1755-1315/45/1/012003

During the research into the interaction between the power support units and host rock we used the following evaluation criteria [1]:

- initial setting load, the finite resistance, the speed of hydraulic props reaction to the weights for a cycle;
- specific pressure of the support on the roof and the floor, the pressure of the relief valve action;
- rate of the longwall face supporting;
- ratio of the hydraulic sliding;
- roof lagging ratio;
- maximum distance from the face to the front edge of the flap;
- longitudinal and transverse stability of the support unit.

According to the diagrams of pressure records in the head end of hydraulic props we evaluated the modes of hydraulic props operation. The pressure in hydraulic props was recorded after the shift of the complex from the installation chamber to a distance of 30m. Figure 1 presents an example of the compacted recording diagrams of pressure in the hydraulic prop of the powered support unit at the end of the winning cycle of extraction panel 21-1-5 (unit No. 53, upper front).

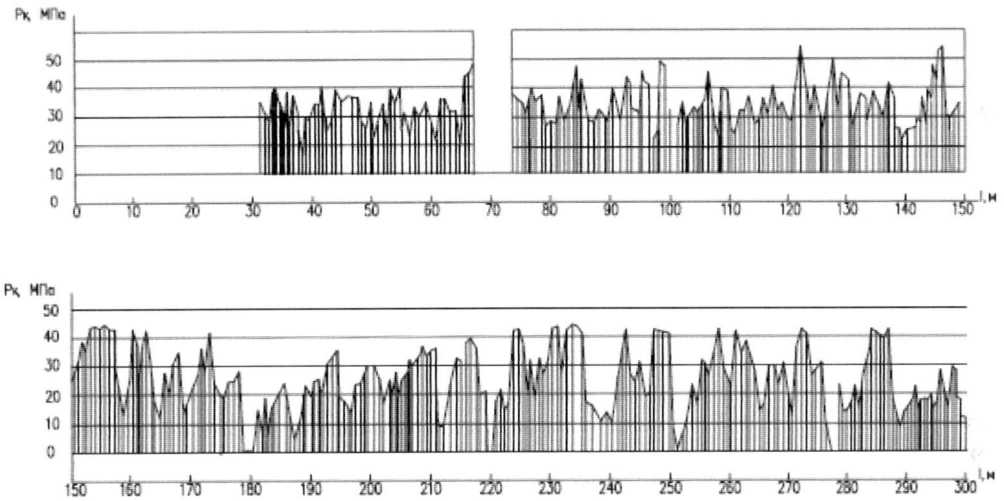

Figure 1. The pressure in the hydraulic prop of a powered support unit at the end of the winning cycle of extraction panel 21-1-5 (unit No. 53, upper front): P_K – setting load pressure, MPa, l – longwall length, m.

The pressure diagrams are compacted with the compaction step of 10 m. The pressure axis is divided into six sections vertically with a scale division value of 10 MPa. The pressure of safety valves opening of hydraulic props is approximately 42 MPa. The analysis shows that the roof supports work in the mode of growing resistance. Working in constant resistance mode occurs during the main roof settling.

The diagrams of the hydraulic props pressure serve to identify the areas in the extraction panel with varying intensity of rock pressure.

Figure 2 shows the fragment of the pressure record in the supports before the fall of the main roof rock at a distance of 40-50 m from the installation chamber. The hydraulic props worked in this area in the mode of a setting load. In continuation of the winning cycle the hydraulic props worked without weights. The setting load pressure ensured stable operation of the support units. The setting load value varied from 20 MPa to 25 MPa and reached 47-59% of the safety valve opening pressure (the degree of the support nominal resistance was 47-59%) [58].

IOP Conf. Series: Earth and Environmental Science **45** (2016) 012003 doi:10.1088/1755-1315/45/1/012003

Figures 2-3 show the fragment of the pressure recording in the hydraulic props during the main roof settling within the thickness of the active roof.

Figure 2. Fragment of the pressure recording in the front upper powered prop before the first roof caving.

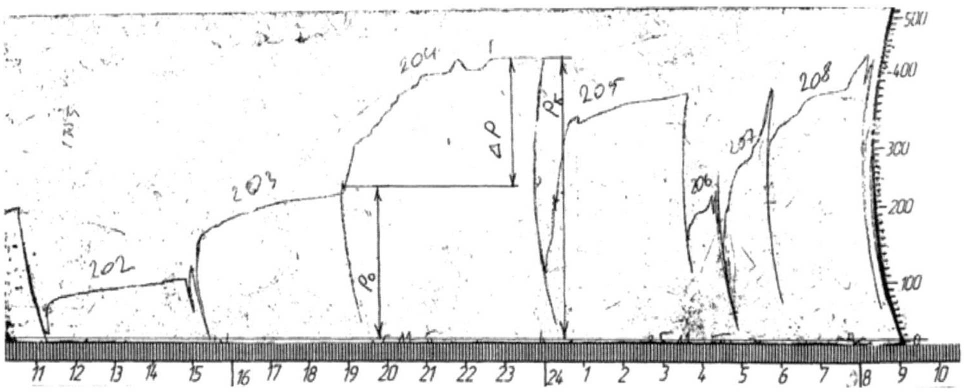

Figure 3. Fragment of the pressure recording in the front upper powered support. The safety valve release (159-163 m from the installation chamber).

The longwall face shift from the installation chamber was 145-163 m. In this area the support props after having been moved and stressed increased pressure at a high speed and operated in the constant resistance mode. Safety valves trigged, props yield occurred. Safety valves released at a pressure of 42-43MPa. There were no "hard" props fitting.

Face stoppage results in the increased load on the support units. Longwall face 21-1-5 in this area of the panel (463 m from the installation chamber) was in the state of emergency. The props worked for a long time under heavy loads and in the mode of constant resistance. The safety valves opened and closed periodically providing constant resistance to displacing rocks. The value of hydraulic prop reaction was approximately 2000 kN.

During the period following the regular roof settling of the rock pressure is insignificant, the rock displacement is balanced by the props reactions, the value of which is determined by the level of the set load.

During secondary settling, when the rock caving occurs within the thickness of the active roof, the props operate in the mode of constant resistance. The pressure in the props varies from 35 MPa to 43 MPa, the props reaction is 1750-2150 kN. The reaction of the support unit without taking into account the unevenly distributed load between the props is 7000-8600 kN.

The histogram of the distribution of the winning cycle duration is shown in Figure 4.

The pressure of the initial setting load in the head end of hydraulic props varied from 2.5 to 29.5 MPa with an average value 14.33 MPa (Figure 5). Load force varied from 125 kN to 1475 kN with an average value 716-718 kN.

The nominal force value of the initial setting load for hydraulic props of the roof support ZF 8000/22/35 is 1546 kN with the setting load ratio equalling 0.77. The extent to which the initial load can be used by average values is:

$$\frac{P_{i.mean}}{P_{i.nom}} = \frac{716}{1546} = 0.46$$

The initial load of the support hydraulic props is used only to 46%, while the nominal setting load coefficient is 77% of the nominal props resistance.

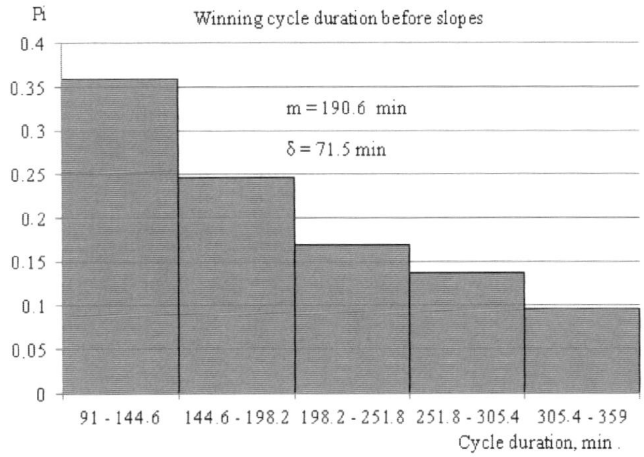

Figure 4. Distribution of the winning cycle duration of the combine in longwall face 21-1-5.

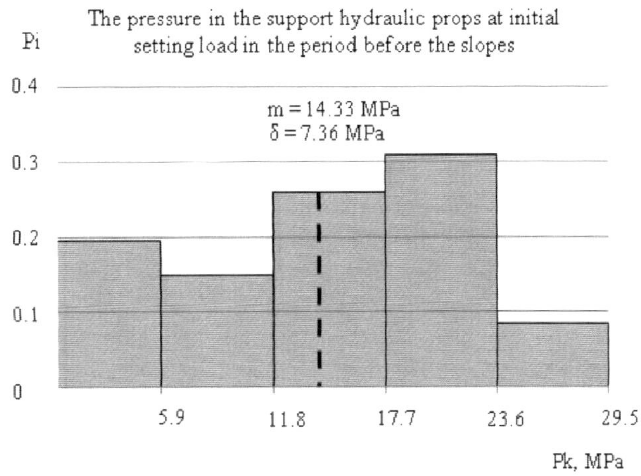

Figure 5. Distribution of the initial setting load of support hydraulic props in longwall face 21-1-5.

Figure 6 presents the histogram of the pressure distribution in the support hydraulic props at the end of the winning cycle, and the actual distribution of the hydraulic prop reaction at the end of the winning cycle is in Figure 7.

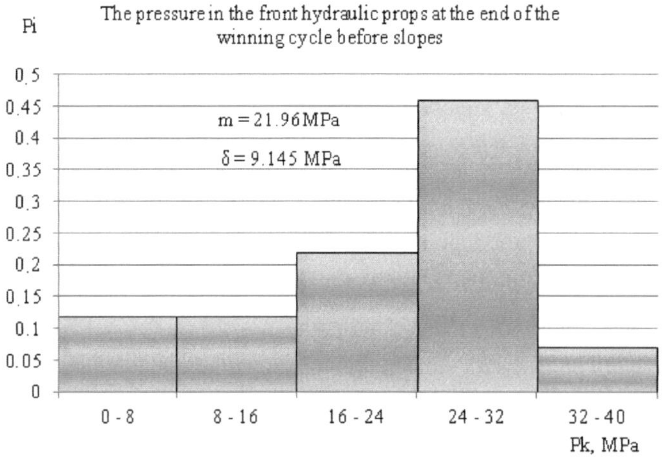

Figure 6. Pressure distribution in the head end of the hydraulic prop at the end of the winning cycle in longwall face 21-1-5.

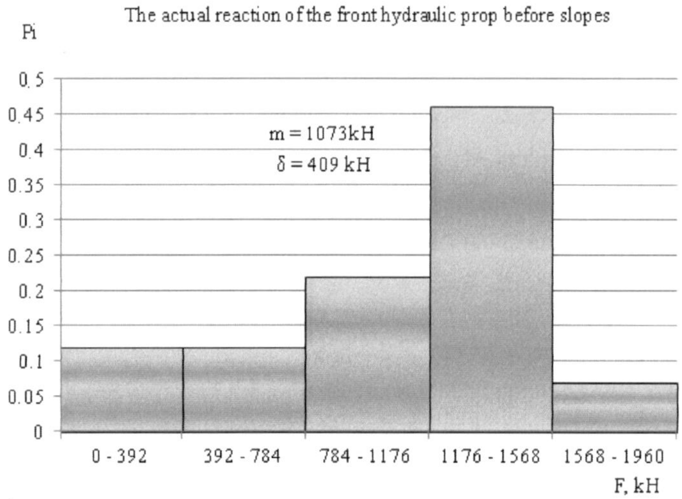

Figure 7. Distribution of the hydraulic prop reaction at the end of the winning cycle in longwall face. 21-1-5.

According to the measurement results shown in Figures 7-8, it is found out that the final response of hydraulic props as well as the initial setting load vary within a wide range, reaching a nominal value of 1960-2000 kN. The average pressure value in the head end is changed from the value of the initial load to the pressure of the relief valve action (42 MPa). The actual average reaction of a hydraulic prop in the observation period was 1073-1100 kN.

The props reach the nominal resistance, safety valves are activated at a pressure of 42 MPa, but at the average actual reaction the degree of the use of the working (nominal) support resistance does not exceed 55%,

$$\frac{R_{fact}}{R_{nom}} = \frac{1100}{2000} = 0.55.$$

The actual value of the coefficient of the initial setting load at the average value of props reaction 1100 kN and the average value of the initial load 716 kN is 0.65 at the nominal value of 0.77.

The results of the performed pressure measurements in the head ends of hydraulic props show that:

- the actual initial setting load is fixed within the range from 2.5 to 29.5 MPa at the average value 14.33 MPa, 46% of the initial load value is used;
- the coefficient of the initial load is 65% and is less than the nominal value of 77%;
- the nominal resistance of the support at use is 55% of the mean value.

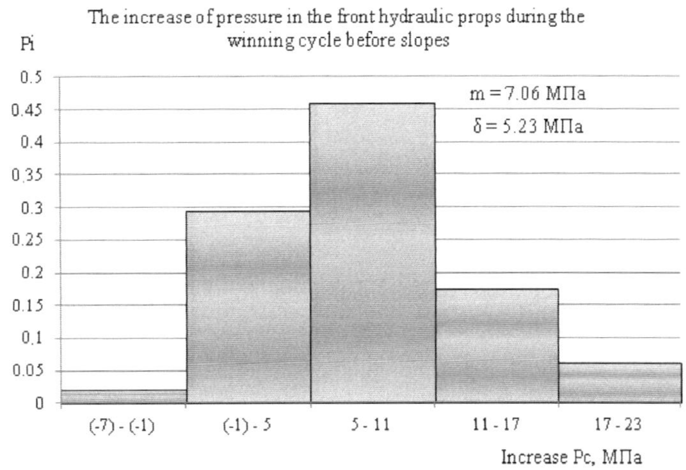

Figure 8. Variation of the pressure increase in the hydraulic prop per winning cycle.

Figures 8-9 show the pressure distribution histograms in head ends of hydraulic props and the rate of the pressure increase that help to assess the nature and intensity of the support props loading.

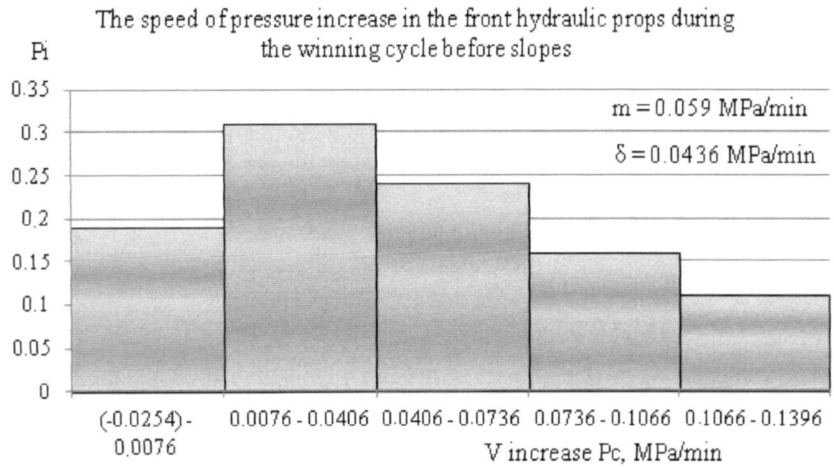

Figure 9. Distribution of pressure increase in the hydraulic prop per winning cycle in longwall face 21-1-5.

The pressure increase in the hydraulic prop was determined by the formula

$$\Delta P = P_f - P_i,$$

where P_f – the finite pressure in the hydraulic prop per winning cycle; P_i – pressure of the initial load in the hydraulic prop per the winning cycle.

The analysis of the histogram in Figure 9 shows that the maximum pressure increase was 23 MPa. Basically, the pressure increase varied from 5 MPa to 11 MPa with a mean value 7.06 MPa. Maximum loads occur during the primary and secondary settling of the active roof. The impact of coal draw on the pressure increase in the hydraulic props is not recorded; the main influence on the pressure increase was produced by the coal cutter.

The speed of the pressure increase during the winning cycle was determined by the formula

$$\upsilon = \frac{\Delta P}{\tau},$$

where τ – the duration of the technological coal mining cycle, min.

The rate of pressure increase changes from 0.0076 to 0.14 MPa/min. The average rate of the pressure increase was 0.059 MPa/min. At a mean initial setting load 14.33 MPa the time of reaching the nominal resistance would be

$$\tau = \frac{P_{nom} - P_i}{\upsilon} = \frac{42 - 14.33}{0.059} = 469\,\text{min},$$

that considerably exceeds the length of the winning cycle (190.6 min). Consequently, according to the mean values the probability of reaching the nominal resistance by the support props is insignificant.

Insignificant pressure rises appear with a high frequency, having no threat to hydraulic props. Pressure rises leading to the nominal resistance in the mode of constant resistance appear at a much slower rate, total pressure in the hydraulic props is not beyond the limits set for pressure relief valve (42-43MPa). The rate of pressure rise is insignificant; safety valves are activated without delay, excluding the "casting" of pressure in the head ends of hydraulic props. The time of reaching the nominal resistance is 85-86 minutes, which is less than the average duration of the technological coal extraction cycle (190.6 min). Consequently, it is possible for the props to work in the constant resistance mode, but the transition to the constant resistance mode is not dangerous, because the rate of pressure increases is negligible (0.14 MPa/min), the safety valves are activated without delay.

It is established that:

- pressure increases in the props are highly frequent, insignificant in value and present no danger for the hydraulic props;
- pressure increases exceeding the nominal value (pressure setting of relief valves) is not observed;
- limit pressures in the hydraulic props causing safety valves activation are improbable.

The number of cycles, in which the safety valves were activated during mining the extraction pillar 21-1-5, amounted to 5.4% of the total winning cycles.

To assess the interaction between powered support units and host rocks it is important to consider the indicator of the uneven distribution of pressure acting on the unit between hydraulic props. Table 1 summarizes the data on the pressure distribution between the unit props in the process of mining the extraction pillar 21-1-5 during the secondary settling of the active seam roof.

The analysis of the data in Table 1 shows that in this sector relatively the same initial load on hydraulic props was set, the ratio of non-uniformity between the face props was 0.97, between the goaf props – 1. The front row props were loaded with an average force of 1041 kN, the back row – 967 kN.

Table 1. The uneven distribution of pressure between the support unit hydraulic props.

Face Row			Goaf Row		
Right (low), (MPa)	Left, (MPa)	K_H	Right, (MPa)	Left, (MPa)	K_H
Initial Load					
20.55	21.09	0.97	19.5	19.2	1
Finite Reaction					
Mean Values					
21.36	20.45	1.04	18.2	22.1	0.82
Maximum Values					
33	36.58	0.9	34.9	35.6	0.98

The reaction of the front (face) hydraulic props by mean values differed little from the reaction of the goaf hydraulic props. In the face row the props reaction was 1068 kN and 1022 kN, the coefficient of variation between the reactions was 1.04, in the goaf row respectively 910 kN and 1105 kN, the coefficient of variation between the props was 0.82. According to the maximum values the reaction of the face row was 1650 kN and 1820 kN, the coefficient of variation – 0.9, the goaf props respectively 1745 kN and 1780 kN, coefficient of variation – 0.98.

The reactions of hydraulic props in the face and goaf rows are not significantly different. The hydraulic props of a support unit are loaded almost similarly.

The evaluation of hydraulic props loading parameters results in the following conclusions:

- powered support parameters are used insufficiently, the initial load – 46% of the nominal, hydraulic props response – 55%;
- the actual initial load ratio is 0.65 at a nominal value of 0.77;
- the support hydraulic props most of the time operate in the mode of the specified load and the growing resistance, without reaching the nominal resistance level.

The constant resistance mode is achieved during periods of primary and secondary settling of the active roof. The share of work of hydraulic props in the constant resistance mode during the extraction pillar mining on the area of 77 0 m was 5.4% of the total winning cycles;

- pressure increases in the support hydraulic props occur smoothly at a low speed. The average rate of pressure increase is 0.059 MPa/min, maximum – 0.14 MPa/min. Sudden roof settlings were not registered, pressure "castings" in the head ends of hydraulic props during transition into the mode of operation with a constant resistance are absent;
- the mean time of achieving constant resistance mode is 469 minutes and is much longer than the average duration of the winning cycle – 190.6 min, there is no possibility for hydraulic props to work with constant resistance.

The time of achieving the constant resistance mode during the regular settlings within the active roof thickness is 86 minutes and is much less than the duration of the winning cycle. The possibility of hydraulic props to operate at a constant resistance is high.

The initial load for the support hydraulic props of the face and goaf rows is specified with the unevenness ratio 0.97-1.0, while the load is distributed uniformly between the hydraulic props, unevenness ratio varies in the range 0.9-0.98. Coal sloughing occurs periodically, mainly during regular periods of active roof settling. The probability of coal sloughing occurrence at various locations along the length of the longwall face is presented as a histogram (Figure 10).

The analysis of the data presented in Figure 10 shows that the coal sloughing in the longwall face appears at the head- and tailgates, in some local areas of the longwall face and along the entire length of the face simultaneously.

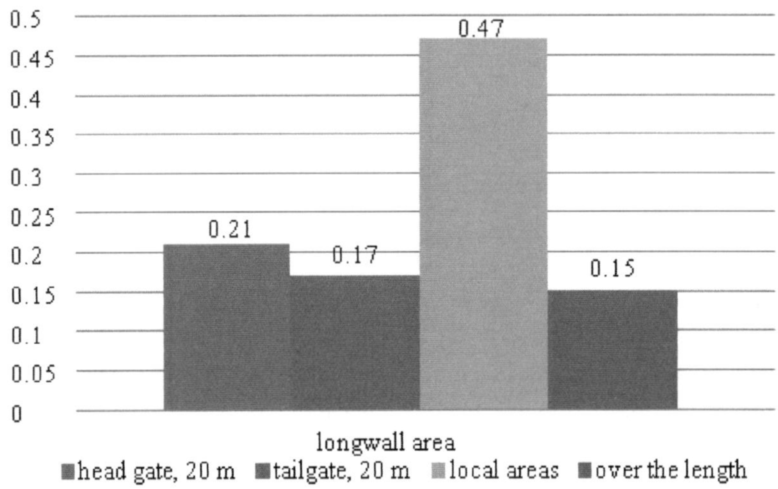

Figure 10. The probability of occurrence of coal sloughing at various locations along the longwall face.

The most frequent occurrence of coal sloughing is observed in local areas, the length of local areas of coal sloughing varies from 1 m to 16 m, the mean value of the local area length is 7.8 m (Figure 11). At head- and tailgates the coal sloughing is less common than in the local areas. Coal sloughing over the longwall length occurred during the main roof settling, i.e. coal sloughing is a periodicity function of the roof settlings, it significantly affects geomechanical parameters, reduces the coal strength in the seam edge, displaces the maximum of the bearing pressure deep into the solid. The observations show that the maximum of the bearing pressure is displaced 4.5-8 m deep into the solid from the face line.

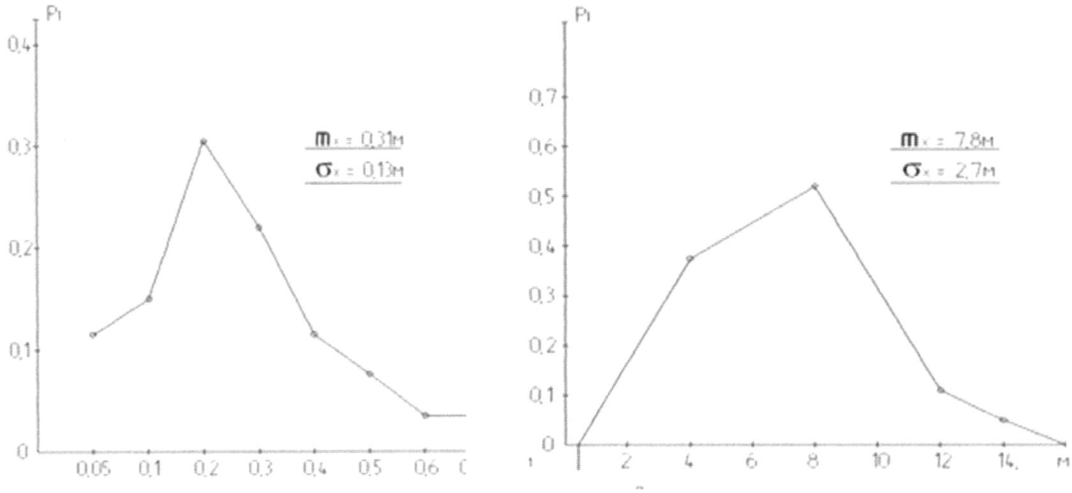

Figure 11. Coal sloughing parameters in longwall face 21-1-5: a – sloughing depth; b – the length of local areas where coal sloughing occurs.

3. Conclusions

The maximum distances from the face edge to the front end of the support vary widely depending on the condition of the face and top-coal. According to the measurements the spacing varied from 0.26 m to 0.42 m, the mean value is 0.4 m.

Coal sloughings and falls from the top-coal through the unit-to-unit gaps occurred mainly during the move of the support units. According to the degree of coal spillage the astel supporting ratio is estimated at the level of 0.9. Unit-to-unit gaps changed in the range from 2 mm to 30 mm, the mean value is 4 mm. It should be noted that these gaps exclude coal spillage in a static state of the support units, the coal spillage occurs when support units are moved. Face protective shields prevent coal falls and sloughings in the face if coal sloughings are not more than 0.3 m. If the coal sloughing value is higher the protective shields are ineffective.

References

[1] Rzhevskiy V V and Novik G Y 1978 *Fundamentals of Rock Physics* (M.: Nedra) p 390
[2] Ermakov A J et al 2009 *Proc.of the II Int. Scientific and Practical Conf. on Innovations – the Basis of the Integrated Development of Coal Industry in the Russian regions and CIS Countries* (Prokopyevsk: KuzSTU branch) pp 90–95

Ventilation of mines developed by the combined method of coal mining

Val V Senkus[1], A Yu Ermakov[2] and V V Senkus[2]

[1] LLC "Proektgidrougol-N", 21 Bardina Ave, Novokuznetsk, 654041, Russia
[2] LLC "Siberian Scientific Research Institute of Coal Processing", 1 Gornaya Street, Prokopyevsk, 653000, Russia

E-mail: senkusvv@suek.ru

Abstract. The paper considers the features of ventilation of mines which are developed by the combined method of coal mining. It also provides recommendations for placing the flank and central ventilation holes while mining flat and steep seams from open pit sides, as well as anticlinal and synclinal deposits.

1. Introduction

Currently, mining of coal deposits by combined method is rare, there is no methodological framework for the flowsheets design of ventilation, drainage, opening, seams development, etc. Design of a common ventilation flowsheet for open and underground mine workings consists in the formation of air flows and their control to ensure standard indicators of air quality in all areas of a mine and open pit.

The solution of the problem consists in the constant replacement of the air, contaminated because of mining operations, by fresh air flows, and can be achieved by using the natural laws of aerodynamic processes in the atmosphere of the Earth in combination with the sanitation of air flows in mines and providing the necessary speed of movement through the artificial creation of difference of atmospheric pressures at the beginning and the end of the specified interval of space.

Ventilation of open and underground workings, which are developed by the combined mining method, conforms to the general principles, reflects the specific combination of technologies and their interaction.

2. Typical features of ventilation in the mines developed by the combined method

Ventilation of mines, which are developed by the combined mining method, has its own specific features, the essence of which lies in the fact that underground mining is carried out from the sides or from the bottom of open pit, being a work area in open pit mining, where the air is full of dust, exhaust gases of the vehicles and rock gas, exceeding the MPC, and its airing is carried out by mixing air masses in the atmosphere and is restricted by the basin-like character of the man-made relief and weather conditions.

Another problem of ventilation is massive explosions, when the exit gasses are enforced into the mine workings, so the underground operations are stopped for 1-3 days for airing at a depth of 180-250 meters underground.

3. Suggestion

The analysis of technological ventilation flowsheets used in the combined mining method shows that they can be of three types (Figure 1-3).

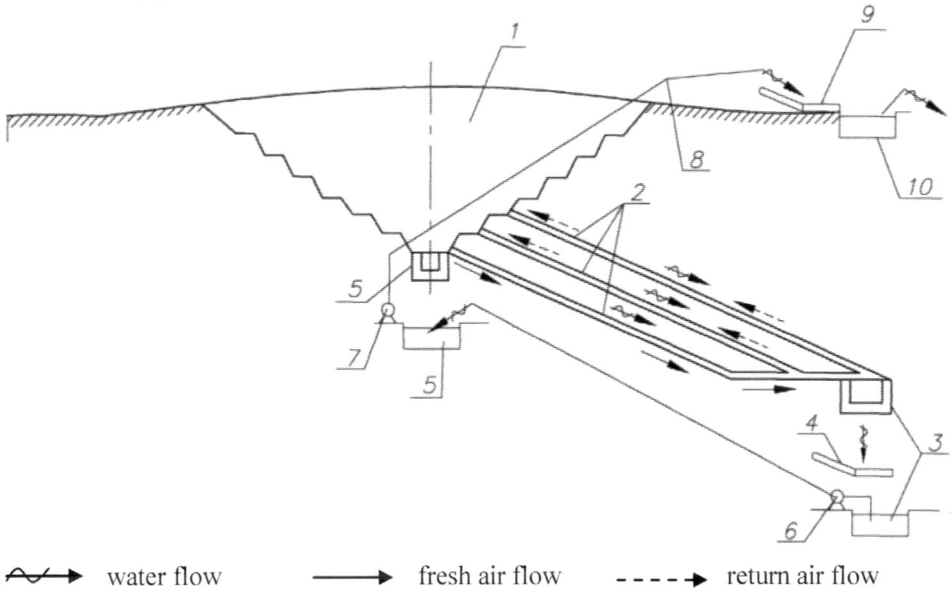

Figure 1. Ventilation of workings in the flat and coal steep seams developed by the combined mining method.

Ventilation of flat and coal steep seams tapped by inclined shafts and mined by the combined method through the seam or rock, depending on the gas content of rock is carried out by the fans of local ventilation or stationary ventilation systems, located in the windward side of the open pit sides along the flexible or rigid air tube. Development of the mine field reserves is traditional for flat and steep seams.

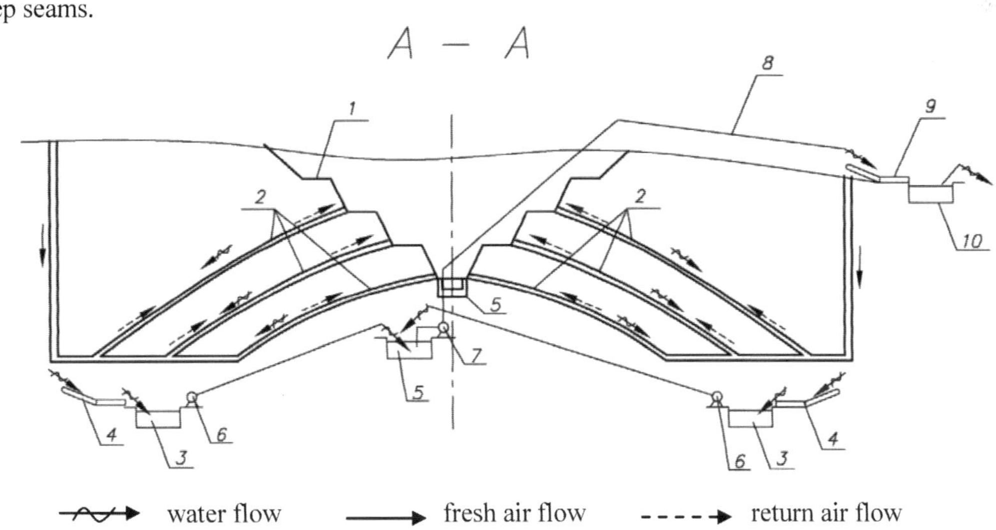

Figure 2. Ventilation of underground workings mined by the combined method of mining of anticlinal coal fields.

For independent ventilation of underground mining, flank (Figure 2) or central (Figure 3) ventilation holes should be used to the lower points of the reserves development, next to which settler ponds of underground drainage are located.

Safety rules do not limit the pressure and the speed of air supply, as the holes are deserted and cased with large-diameter pipes. The location of the holes on the flanks and in the center of the synclinal coal deposit is chosen based on the compatibility conditions with the flowsheets of underground drainage, mine take boundary, the possibility of airing the maximum volume of open-up reserves.

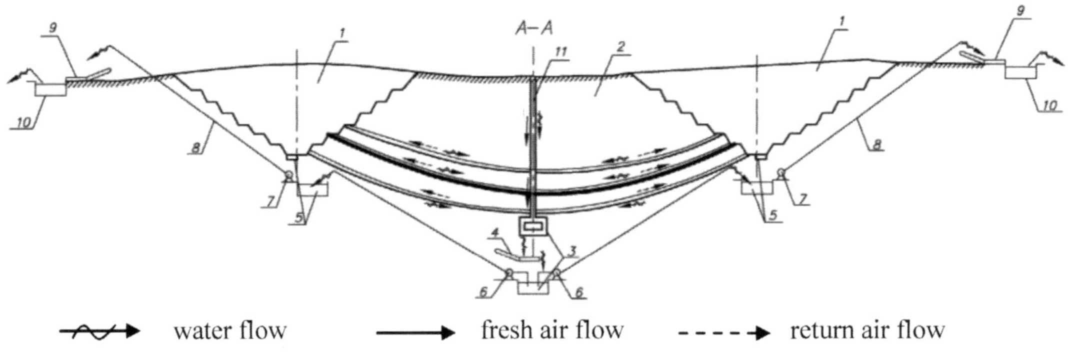

Figure 3. Ventilation of underground workings mined by the combined method of mining of synclinal coal fields.

Ventilation holes can be drilled individually for each seam of the mined suite, but should be connected to one fan unit, which ensures the independence of their airing.

The air from the ventilation holes is distributed over the mine workings by the arrangement of stoppings, gates and shutters.

The return air flow is ejected into the working area of the open pit creating an upcast air into the atmosphere, providing additional ventilation of open workings.

4. Specific features of ventilation in the mines developed by the combined method

The specific features of ventilation of mine workings are the following:

1. The main fan unit should be located outside the contamination zone (Figures 2, 3).
2. The ventilation design should provide the incline ventilation scheme with fresh air stream to the lower point of the mine field and to distribute it to the objects of airing.
3. Placement of the main air fan in the open cast workings should be as farther as possible from the open mining operations, and the main fan location should be chosen, taking into account the wind rose and the climatic conditions of the construction area.
4. Carry out inspections of the width of the barrier pillar between the open and underground mining operations according to the condition of the wind permeability.
5. Ventilation design should not miss the opportunity to direct the return air flow from the mine to ventilate the mined-out space in the open cast and create favorable atmospheric conditions in its working area.

5. Conclusions

The offered flowsheets of ventilation allow the combined method of coal mining to be adjusted for complex mining and geological conditions and an effective ventilation of mine workings to be ensured.

Reliability evaluation of the mining flowsheet of seam 21 in "Olzherasskaya-Novaya" mine LLC "UC Yuzhny Kuzbass"

A Yu Ermakov

LLC "Siberian Scientific Research Institute of Coal Processing", 1 Gornaya Street, Prokopyevsk, 653000, Russia

E-mail: ermakovay@suek.ru

Abstract. The paper assesses the flowsheet of mining seam 21 at "Olzherasskaya-Novaya" mine; based on the experience gained while testing the technology and calculating the reliability parameters the mechanisms, which break down most frequently, are specified.

1. Introduction

Reliability of flowsheet is the result of its functional capacity that is determined by different methodologies. To evaluate the reliability of the mining flowsheet of seam 21 in longwall face 21-1-5 we used the following indicators: Run to Fail T_{RF}; failure recovery time T_r; availability factor K_a; failure factor K_{ff}; the failure rate K_{fr}; downtime rate K_{dr} [1].

Evaluation indicators were defined for the process flowsheet and subsystems: the face, conveyor transport, power supply, water supply, ventilation; and for the main face equipment: shearer, powered support, face and goaf conveyors. Run to Fail and failure recovery time were calculated according to the formulas presented in [2].

2. Observation results and their analysis

The results of observations and processing of statistical material helped to identify the number of failures of organizational, technological and technical reasons, causing the longwall face stoppage.

During the test period of 13 months there were 776 failures, most of which can be attributed to the slope belt conveyor, gas protection equipment, shearer, face and goaf conveyors. Failure rate for the slope conveyor is 0.215, gas protection equipment – 0.137, shearer – 0.126, face conveyor – 0.088, the goaf conveyor – 0.067.

Frequent stops of the longwall face occur due to high concentrations of methane and the gas protection activation, that was observed during roof settlings, the joint extraction and release of coal during the periods of working with high face advance rates.

Total downtime is 134515 min. (about 2242 hours, or 93.4 days). The largest proportion of the downtime accounts for the goaf conveyor, the downtime ratio is 0.3. A significant part of downtime is registered for the slope belt conveyor, downtime ratio is 0.104. The share of downtime for gas protection equipment is small, the ratio is 0.034.

On the ground of statistical data, we determined the mean Run to Fail of the complex – T_{RT}, the mean failure recovery time – T_r and the average availability factor of the flowsheet: Run to Fail is T_{RT} = 196 min; failure recovery time T_r = 111 min; availability factor K_a = 0.638; failure ratio K_{fr} = 0.567.

Availability and failure factors were determined by the elements of the flowsheet taking into account long periods of inactivity (more than one day) and excluding them. The availability factor, with regard to long downtime has the least value for the goaf conveyor – 0.839 and for the shearer – 0.866. The greatest failure rate is for the goaf conveyor – 0.192 and the shearer – 0.155.

In general, in the group of the main machinery in equipment complex the indices of operational reliability taking into account long periods of downtime are $K_a = 0.65$, $K_{fr} = 0.46$.

The use of working time of a shift is considered for a 6-hour and 8-hour shift. The downtime and failure rates were calculated and the time distribution in a shift for performance of coal mining operations, for emergencies and technological regulated downtimes.

Figure 1 shows a diagram of the time distribution in a shift according to the main processes (Figure 1a), and the longwall face downtime rates due to equipment failures and other reasons (Figure 1b) for an 8-hour work shift.

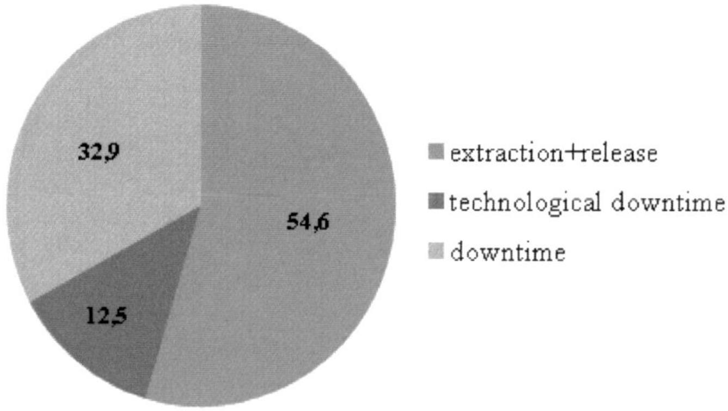

a

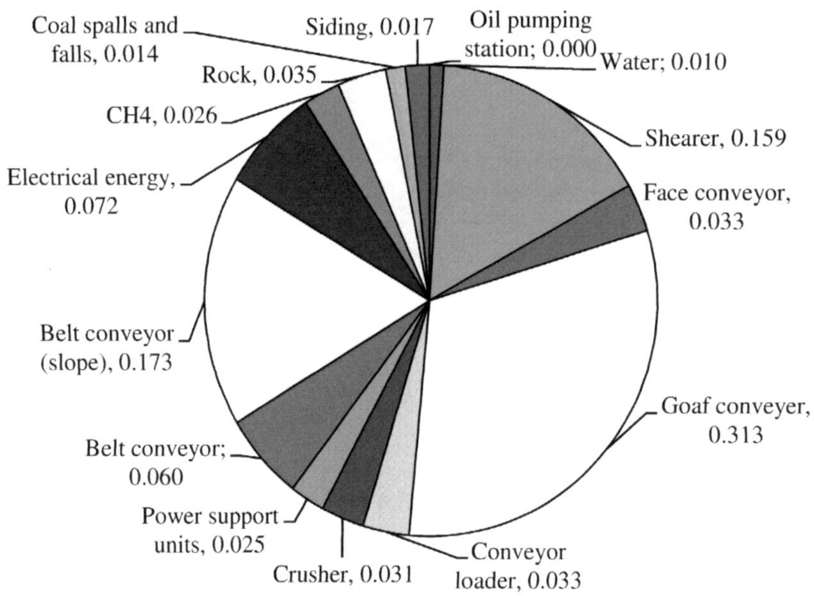

b

Figure 1. Time distribution of 8-hour work shift with long periods of downtime: a – the shift time distribution between main processes; b – downtimes distribution (K_{dr}) by the main factors.

Working time of a shift is distributed in the following way: coal extraction and release make 54.6% of the shift time; downtime accounted 32.9%, technological downtime – 12.5%. The most downtime periods were due to failures of the goaf conveyor and the shearer.

It should be noted that some elements of the equipment require design improvements: insufficient jacks effort in advancing the roof support and the face conveyor, the jack pulling the goaf conveyor, jacks lifting rotary shearer gears; electric control circuit of the goaf and the face conveyors requires modification; the shearer weak points are its cutting drums.

The powered support often fails due to hoses, replacement of which is a time consuming operation.

Figure 2 shows the shift time distribution between the efficient use of the time and the downtimes in a 6-hour work shift.

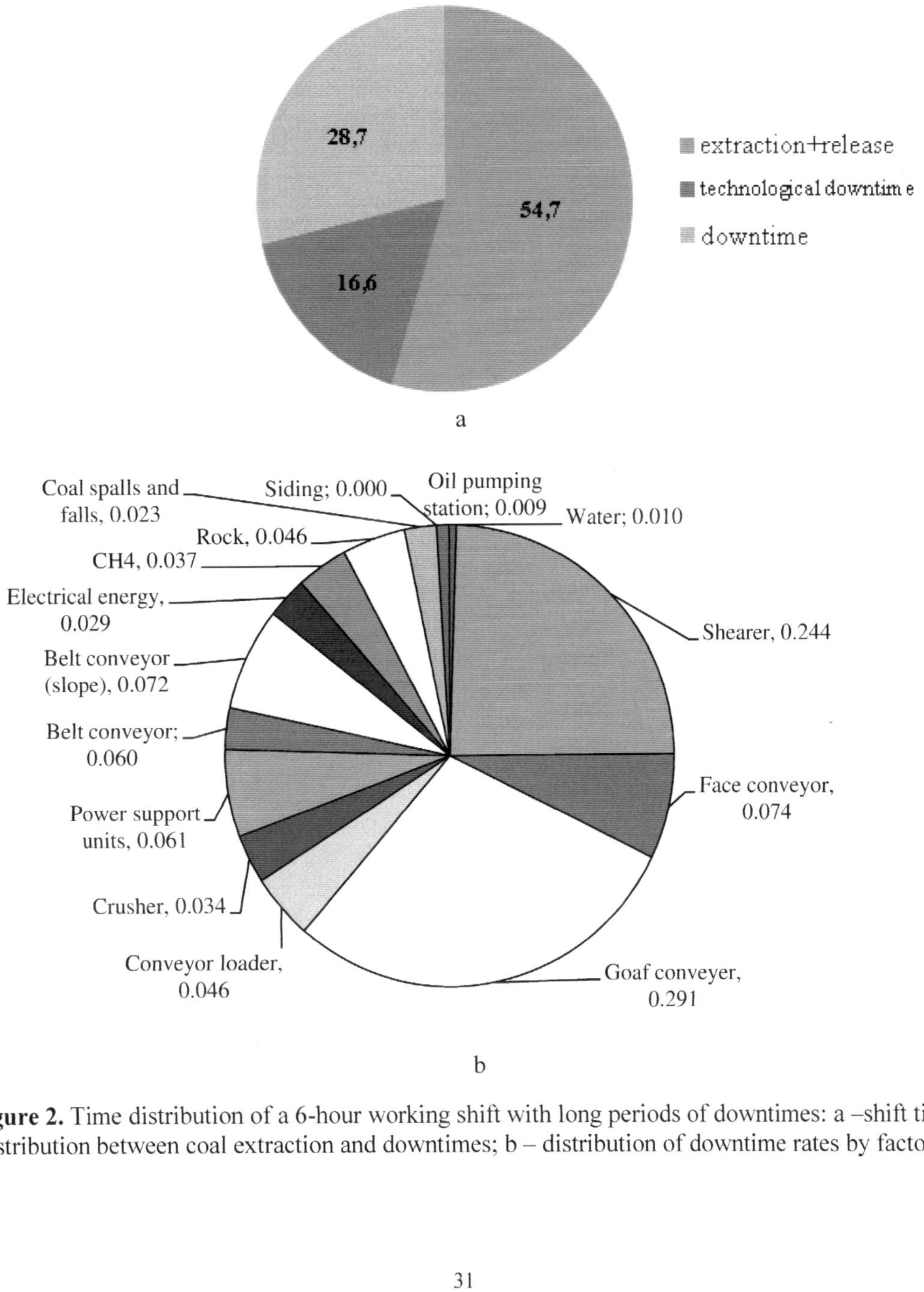

Figure 2. Time distribution of a 6-hour working shift with long periods of downtimes: a –shift time distribution between coal extraction and downtimes; b – distribution of downtime rates by factors.

3. Conclusions

The working time of a 6-hour shift is distributed as follows: coal extracting and coal release 54.7% of the shift time; overall downtimes 28.7%: technological downtimes 16.6% that characterizes the technological flowsheet as a comprehensive one with regard to the modern technology.

References

[1] Novoseltsev S A, Ermakov A Y et al 2009 *Proc. of the II Int. Scientific and Practical Conf. on Innovations – the Basis of the Integrated Development of the Coal Industry in the Regions of Russia and CIS countries* (KuzGTU Prokopyevsk branch) pp 131–137

[2] *Methodology of Operational Testing the Mining Technology in Seam 21 at "Olzherasskaya" mine of "Olzherasskii Openpit" by Top-coal drawing* 2006 (Prokopevsk) p 45

A study on directions of significant efficiency increase of rock fracture by tools equipped with super hard inserts from composite

L T Dvornikov[1], V I Klishin[2], S M Nikitenko[2], V A Korneyev[1] and P A Korneyev[1]

[1] Siberian State Industrial University, 42 Kirova Street, Novokuznetsk, 654007, Russia
[2] Federal Research Centre of Coal and Coal Chemistry of Siberian Branch of the Russian Academy of Sciences, 10 Leningradsky Ave, Kemerovo, Russia, 650065

E-mail: nsm.nis@mail.ru

Abstract. In the paper the directions of a significant increase in effectiveness of mine rocks destruction by tools equipped with super hard inserts from composite materials are reviewed and justified. Designs of mining drill bits with the cutting insert in the form of elliptical Cassinian oval and the asymmetric ring cleaves are suggested. Versions of laboratory stand constructions are developed in order to determine the power consumption of rock destruction.

1. Introduction
In mining industry major part of finances is used to arrange the works associated with crushing the rocks. During mechanical impact on rocks with special tools in order to crush it the tool itself also experiences high loads depending on physics-mechanical rock properties [1]. Two tasks become actual in this situation – provision of a significantly much higher tool strength compared to the strength of rocks, and creation of such tool forms that would make the crushing process less power consuming.

The most demanded rock crushing tool used at present at many coal and ore mines is a cutting bit of drilling machines and coal cutting machines: radial, tangential, rotating, blades for rotary drilling and also bladeless drilling crown bits [2]. Search for scientific and technical solutions to improve the efficiency of rock destruction with mining tool is an actual task for mining industry. The implementation of such studies in recent years is carried out by the Institute of Coal RAS SB within FCP theme: "Development of experimental combined tool designs using the super hard composite materials for the efficient rock destruction". The solution of the problem in the ongoing studies is expected to be achieved by means of theoretical justification of samples design of the combined instrument on the basis of reinforcing inserts from super hard composite materials with the further study of their interaction with the rock on specially designed laboratory benches.

2. Research into the problem
Efficiency increase of rock crushing by mining tool can be achieved by different means: changes in tool geometry, use of new more productive and energy efficient modes of rock crushing, as well as the use of new materials in the manufacture of cutting inserts [3]. The first direction includes research on the development of various types of mining tools structures, providing guaranteed turnability of revolving cutters, prevention of rear cutter face from lowering, more rational armoring of bladeless

tool, as well as expansion of application range of radial cutters according to the rocks strength without changing the cutting insert geometry.

Based on the results of the fulfilled researches new designs of drill bits are developed: with a cutting insert in the shape of Cassinian elliptical oval [4] and also with asymmetric ring cleave [5]. The cutter with a cutting insert of elliptical Cassinian oval shape is shown in Figure 1a. It consists of a body 1, two wings with hard alloy inserts 2 of elliptical Cassinian oval shape, blades of which 3 are set along the big oval axis, wherein the big axes of the ovals are inclined to the tool axis at an acute angle α (Figure 1a).

One of the properties of elliptical Cassinian oval is the presence of two points with a curvature tending to infinity, which allows the full contact to be achieved when interacting with the face at these points. This feature of the cutter helps to reduce cutter blades blunting 3, in contrast to the existing cutters, in which the interaction of the blades with the face is carried out along the stress concentrators placed on them.

In order to reduce the energy consumption of rock drilling, by providing their destruction with large cut-offs, within the frames of the study the drilling bit with asymmetric ring cleave was designed (Figure 1b). The drilling bit was created on the basis of RKS bit design, consisting of a body and discontinuous blades placed symmetrically to the bit axis [6]. The disadvantage of the RKS tool is the fact that both inner semi-blades cut on the face of the borehole concentric strips, easily imposed on each other, which does not lead to active face destruction between the cutting strips, i.e. power consumption of destruction process is high.

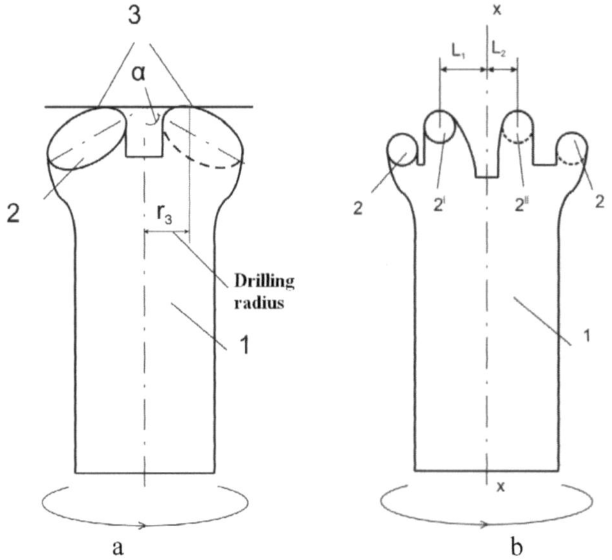

Figure 1. Drilling bits with the cutting insert of elliptical Cassinian oval shape (a) and asymmetric ring cleave (b).

Considering this drawback of PKS cutter, in the design of the cutter with asymmetric ring cleave the inner semi-blades are located asymmetrically with respect to the axis of the cutter rotation (Figure 1b). The developed cutter design with asymmetrical ring cleave allows on its basis the cutters with different semi-blade combinations to be created corresponding to any given borehole drilling conditions.

Drilling tool with asymmetrical ring cleave (Figure 1b) consists of a body 1, two wings with discontinuous blades designed as peripheral 2 and the internal 2' and 2" semi-blades, while internal semi-blades 2' and 2" are set asymmetrically with respect to cutter axis, at a distance L_1 and L_2. During drilling, the cutter is rotated according to the arrow in Figure 1b with simultaneous axial advance to the face. Semi-blades 2, 2' and 2" start cutting rock from the face. Due to the asymmetric installation of internal semi-blades 2' and 2" the cut strips overlap with each other, thus, providing face destruction of the drilled borehole with big pieces, that results in reduction of power consumption of the drilling process.

As it was previously noted, when considering ways to improve the efficiency of rocks destruction by drilling cutters, one of the most urgent problems to be solved during their design process is to prevent the rock shear by the cutter rear side. Theoretical studies carried out in the framework of the Federal Target Program, allowed fundamentally new designs of cutting blades of drilling bits to be justified.

In Figure 2 the bit movement geometry can be seen, performing rotational movement around the axis $0^{I}0$ with speed ω and simultaneous forward advance into the depth h per one revolution.

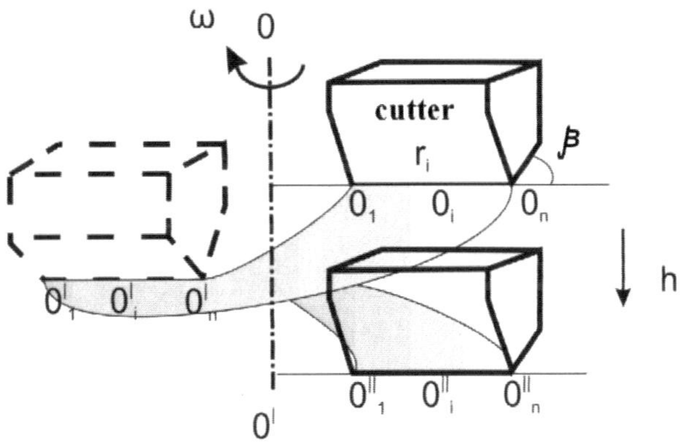

Figure 2. Geometry of drilling bit movement.

From Figure 2 it can be seen that each point of the cutter blade 0_1, 0_i, … 0_n, at distance r_i from the tool axis, passes the path $0_1 0^{I}_1 0^{II}_1$, $0_i 0^{I}_i 0^{II}_i$, … $0_n 0^{I}_n 0^{II}_n$ per one cutter turnover. Helical trajectory scan of each point 0_1, 0_i, … 0_n in Figure 2 represents the hypotenuse of a right triangle, the legs of which are the distances of the drilling tool advance h and the circumference length with r_i radius described with a corresponding point relative to the axis $0^{I}0$ (Figure 3a). The trajectory of each cutter blade point is the angle of inclination α_i with the cutting plane of the cutter at a zero advance (Figure 3a). This angle can be found as:

$$\alpha_i = arctg \frac{h}{2\pi r_i}. \qquad (1)$$

From relation (1) it can be seen that the value of the angle αi is a variable and decreases from the blade inner edge to its outer edge. This fact imposes certain restrictions on the clearance value of rear cutters angle β since achieving equality αi=β drilling cutter stops cutting, causing the rock shear.

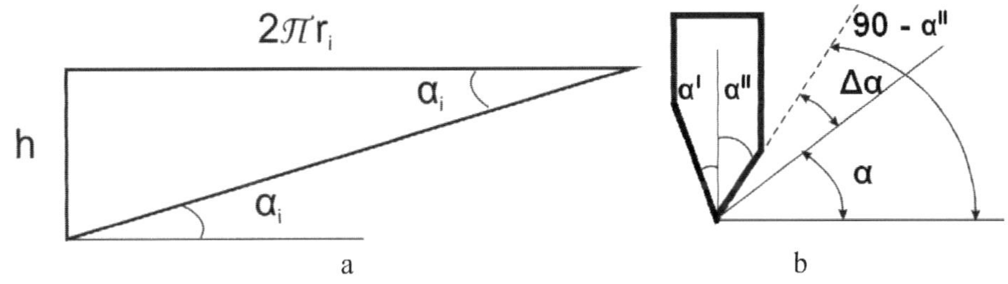

Figure 3. Helical trajectory scan of the cutter point (a) and cutting geometry (b).

Usually, in practice [6], the mining tools rear angle is within 5°÷20°. With positive front angle the rear angle increase leads to weakening of the cutting edge of the tool and may lead to its failure. With negative front angle, as the studies have shown, rear angle can be increased up to 20°÷30°, as the strength of the cutting edge is not reduced, and the wear areas have smaller dimensions.

It is obvious that the the strength of the cutting edge depends directly on the angle of its sharpening which is composed of the front $\alpha^{|}$ and rear $\alpha^{||}$ sharpening angles (Figure 3b). At the same time, to guarantee the prevention from rock shear by the rear cutter blade, the difference Δα between the cutting angle α and the rear sharpening angle should be provided in accordance with condition (2).

$$\Delta\alpha = (90 - \alpha^{||}) - \alpha \quad (2)$$

The cutter design corresponding to condition (2) can be obtained when using variable values of back sharpening angle $\alpha^{||}$ over the entire length of the blade [7]. Considering the plots in Figure 3a and using drilling cutter advance value h, we can calculate the sharpening angle value $\alpha^{||}$ for any point of the blade (Figure 4).

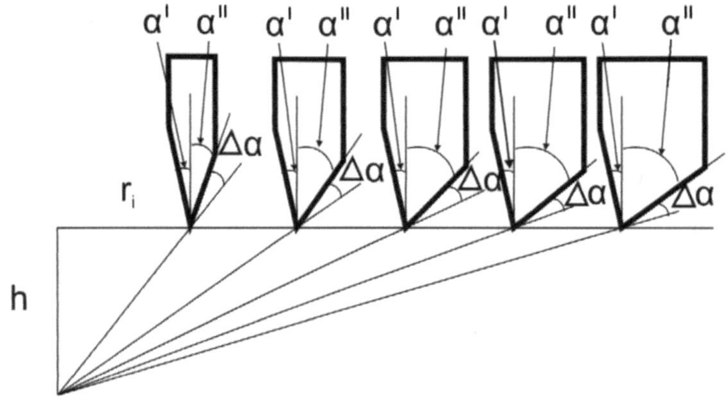

Figure 4. Variation of the rear angle of cutter sharpening from the cleave to periphery.

Thus, the implementation of a variable adjustable sharpening angle $\alpha^{||}$ of the cutting edge across the length of the blade in the design of the tool helps to prevent the rock shear by the cutter rear edge due to the difference Δα between the cutting angle α and rear sharpening angle. Reduction of the sharpening angle from the periphery to cleave can be justified by the fact that the moment of cutting resistance at the points of the tool from the periphery to cleave is reduced because of the reduction of these points radii with respect to the tool axis.

Apart of the considered case, prevention of rock shear by the rear edge of the cutter can be implemented in the design of the drilling bit with a constant blade sharpening angle under the following condition:

$$\alpha^{|} + \alpha^{||} = \alpha^{|||} = const, \quad (3)$$

where $\alpha^{|||}$ – sharpening angle of the bit cutting insert.

Based on the requirements (2) and (3) in order to achieve stable bit operation and to prevent its damage due to insufficient strength of cutting inserts, production of advanced drilling bit models should be implemented with variable front and rear sharpening angles, and in compliance with the cutter sharpening angle constancy [8]. Variation of sharpening angles of cutting insert, satisfying conditions (2) and (3), is shown in Figure 5.

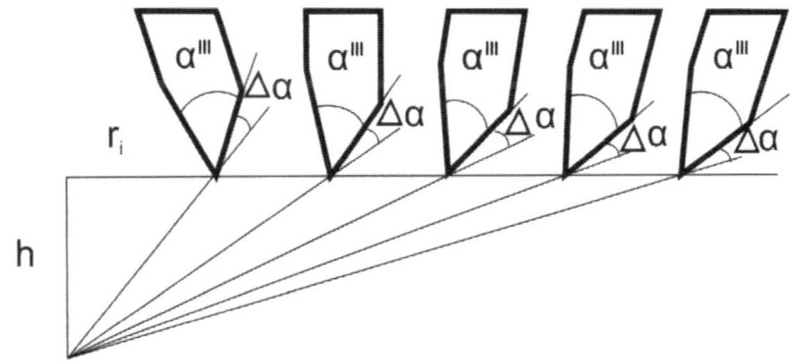

Figure 5. Variation of cutter sharpening angles from the cleave to periphery with a constant sharpening angle.

A three-dimensional view of cutting inserts with variable and constant sharpening angle is shown in Figures 6a and 6b, respectively. Figure 6 depicts: 1 – insert body, 2 – blade, 3 – insert rear edge, x-x geometric axis of the drilled borehole, arrows (a) and (b) show the insert rotation direction with respect to x-x and its advance – the deepening of the cutting insert into the borehole. Points A, B, C, D, are used to indicate the tops of the insert, forming its rear edge, L – thickness of the cutting insert, $\alpha^{|||}$ – the angle of its sharpening.

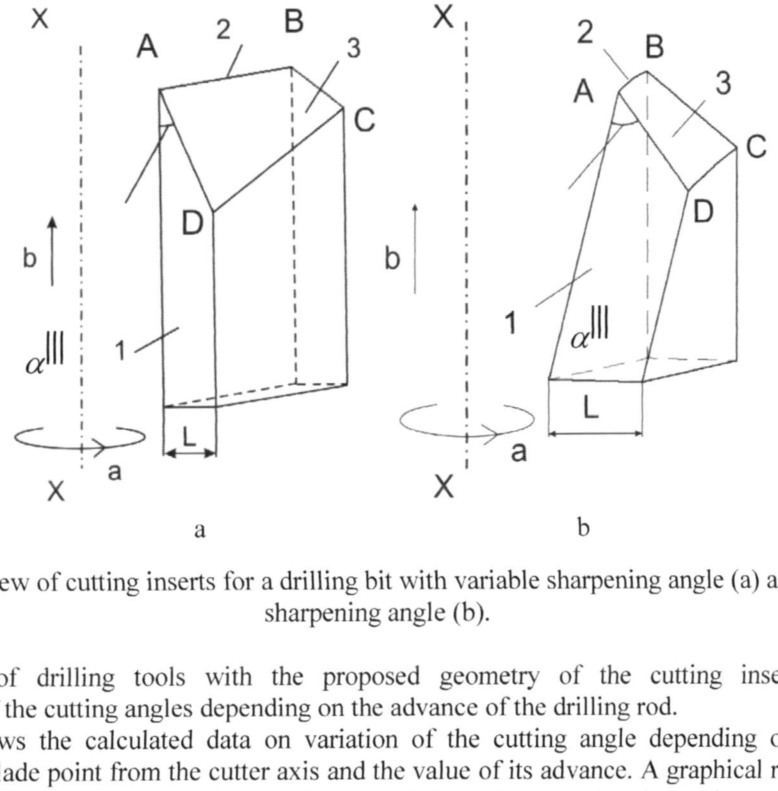

Figure 6. 3D view of cutting inserts for a drilling bit with variable sharpening angle (a) and a constant sharpening angle (b).

Production of drilling tools with the proposed geometry of the cutting inserts, requires consideration of the cutting angles depending on the advance of the drilling rod.

Table 1 shows the calculated data on variation of the cutting angle depending on the radius remoteness of blade point from the cutter axis and the value of its advance. A graphical representation of the calculated data is shown in Figure 7, from which it can be seen that the cutting angle decreases from the inner edge to the periphery, while its values rise as the advance increases.

As it was previously noted the connection between the drilling speed and number of revolutions of the drilling bit is of non-linear character because of the optimal rate of rotation of the cutter for rocks

IOP Conf. Series: Earth and Environmental Science **45** (2016) 012006 doi:10.1088/1755-1315/45/1/012006

with various strength coefficient [6]. Thus, the reserve for improvement of drilling performance may be found only in the advance increase of cutter, resulting in a significant increase in cutting angle (Table 1); and it will cause a decrease in the angle of sharpening and reduction of blade strength with the existing blade manufacturing technology.

Table 1. Variation of a cutting angle depending on the remoteness radius of a blade point from the cutter axis and the value of its advance.

Advance, mm	Radius of remoteness of the blade point, mm								
	5	7	9	11	13	15	17	19	21
3	5.46	3.91	3.04	2.49	2.11	1.83	1.61	1.44	1.30
4	7.26	5.20	4.05	3.32	2.81	2.43	2.15	1.92	1.74
5	9.05	6.49	5.06	4.14	3.51	3.04	2.68	2.40	2.17
6	10.82	7.78	6.06	4.97	4.21	3.65	3.22	2.88	2.61
7	12.57	9.05	7.06	5.79	4.90	4.25	3.75	3.36	3.04
8	14.30	10.32	8.06	6.61	5.60	4.86	4.29	3.84	3.47
9	16.00	11.58	9.05	7.43	6.29	5.46	4.82	4.32	3.91
10	17.67	12.82	10.04	8.24	6.99	6.06	5.35	4.79	4.34

The task of a substantial (in times) increase in drilling productivity, which means the drilling rate, can be expressed as follows:

$$V = h \cdot n, \tag{4}$$

where V – drilling rate, h – advance at one bit rotation, n – number of bit rotations.

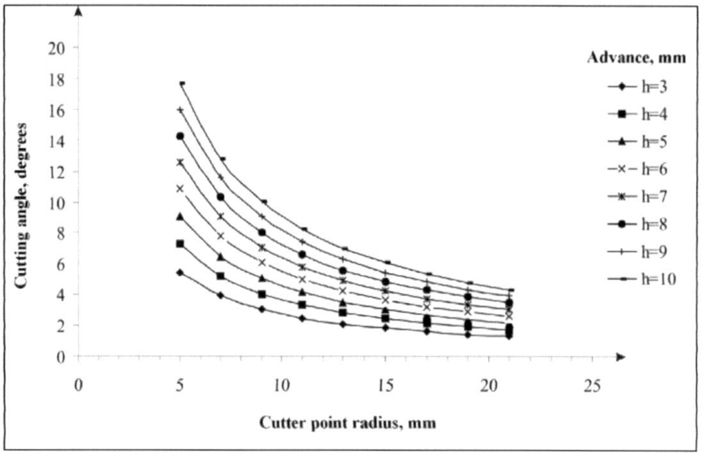

Figure 7. The cutting angle of blade points depending on the distance from the rotation axis of a drilling cutter for different advance values.

Thus, the development of drilling cutters with variable front and rear sharpening angles is one of the possible solutions of the task. The difference φ between cutting angles of points of inner and peripheral parts of the blades depending on the advance value is given in Figure 8.

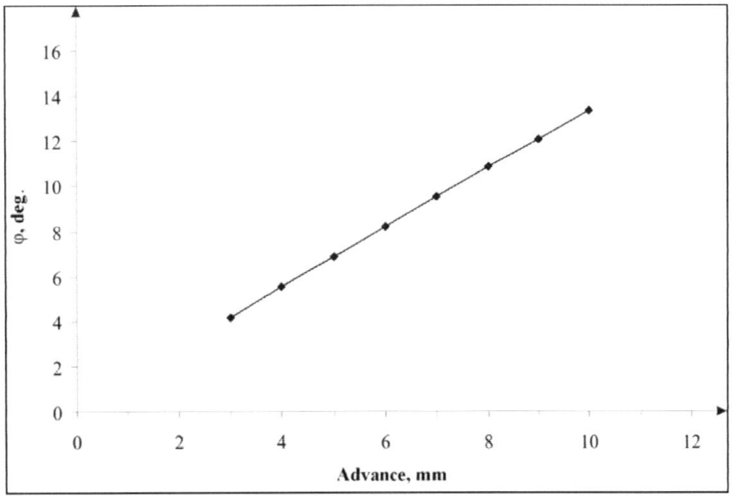

Figure 8. Dependence of cutting angle difference of inner and peripheral parts of a blade on advance value.

3. Results and discussion

Effectiveness evaluation of rock destruction by the developed tool can be carried out with the involvement of various criteria, The basic criterion is the specific power costs for rock destruction. Thus, the choice of a tool design is possible only on the basis of specific information on the lowest possible power consumption for a definite type of rock destruction. In this regard, for the research purposes the methods, which allow the power consumption during the rocks destruction to be determined quickly and with sufficient accuracy, should be used [9].

The method for determination of the power consumption, satisfying all these requirements, and the device for its implementation were developed by L.T. Dvornikov together with N.I. Naumkin [10]. The method consists in selection of samples with irregular shape, measurement of their volume followed by crushing on a hard loading device to a predetermined degree of dispersion with simultaneous removal of fractured particles of specific sizes from the crushing zone.

In accordance with this method, during the fracture of rock samples the pieces punch force and its displacement were registered. As a result the deformation and fracture diagram was recorded. The work used for rock destruction is defined as the area under the curve, and power consumption – as a specific work per initial sample volume unit.

To implement this method for determination of the power consumption several bench modifications were developed with a mechanical and hydraulic punch advancing [11-13]. The laboratory bench for determination of power consumption with a hydraulic punch advancing is shown in Figure 9.

Structurally it consists of a frame 1, perforated cup 2, where the rock samples are arranged, the punch 3 driven by a hydraulic loading cylinder 4 through a manual pump 5. The perforated cup 2 is placed on the frame 1 by means of a support 6, in which a radial-thrust bearing is placed, providing the rotation of the perforated cup 2 with handle 7 to remove the crashed rock.

The energy meter measuring unit consists of DM5007A pressure sensor that registers the force impact of the punch on the rock, and the laser displacement sensor RAS-T5-500-420A, fixing the punch position.

The measured parameters are recorded by the data collection module MSD-200 with the further plotting the deformation and rock destruction diagram by means of a specially developed software for calculation of energy consumption as a specific crashing work [14]. Thus, the work is defined as the area under the graph. A typical form of the graph, built in the software during rock sample testing, is shown in Figure 10.

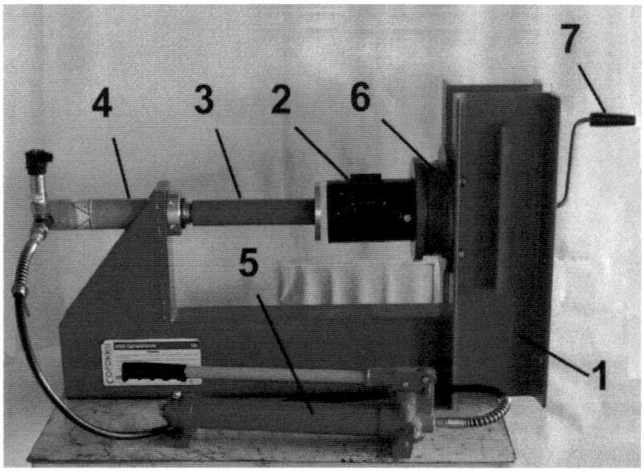

Figure 9. Laboratory bench for determining the power consumption of rock destruction with a hydraulic punch advance.

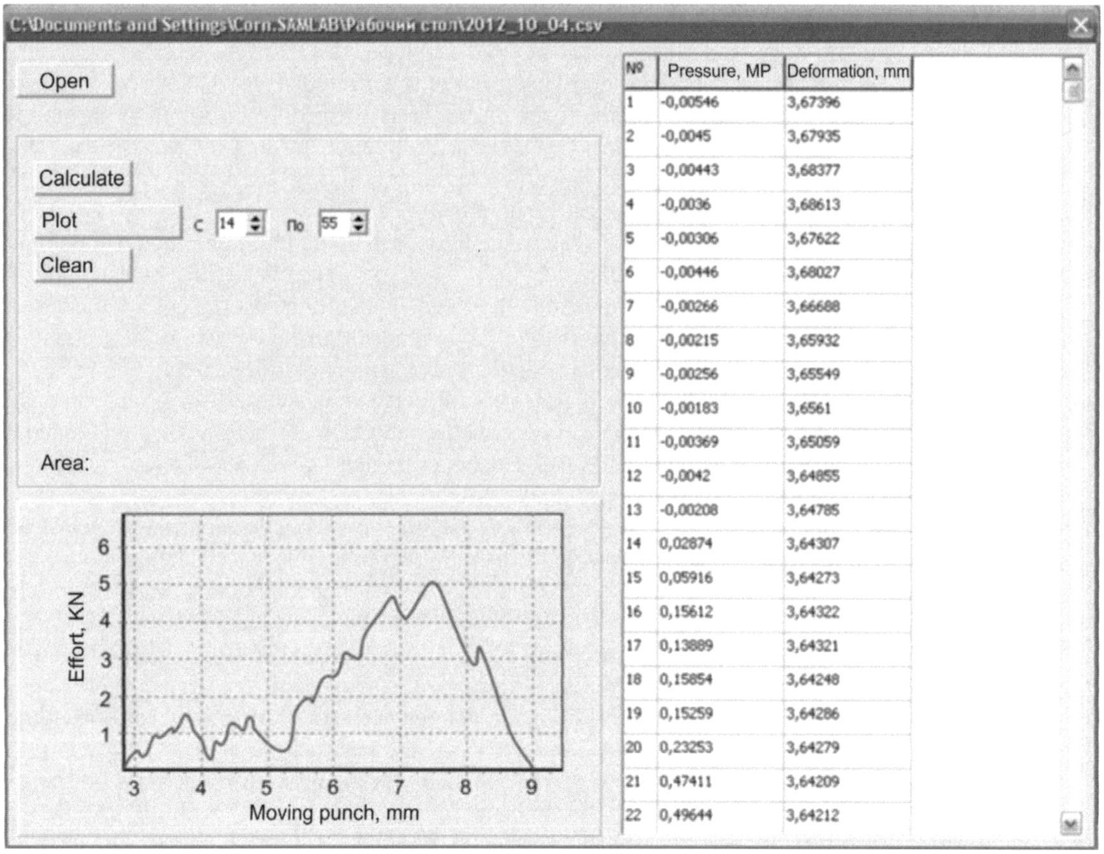

Figure 10. Deformation and rock crushing diagram plotted by the specially developed software.

As noted earlier, the use of new materials for design of cutting inserts is one of the possible ways to improve the efficiency of rock destruction. Theoretical studies at the first stage of FCP have shown that the most promising among them are super hard composite materials on the basis of cubic boron nitride and technical diamond (PCD).

4. Conclusions

With consideration of the conducted research, the recommendations for possible use of super hard composite materials in the design of a new instrument for the effective rock destruction might be as follows:

1. It is more appropriate to use composite materials containing diamond for mining tool design, with preference given to composites with the highest thermal resistance.
2. In some cases, in particular:
   - not high (relatively) hardness of the processed rock;
   - high (from 400 to 900 degrees) temperature in the cutting area;
   - presence of elements in the rock which are catalysts for PDC destruction (iron, nickel, cobalt), possible use of composite materials containing cubic boron nitride.

5. Acknowledgements

The work was financially supported by the Ministry of Education and Science of the Russian Federation within the Federal Targeted program for Research and Development in Priority Areas of Development of the Russian Scientific and Technological Complex for 2014-2020 (agreement No. 14.607.21.0028 from 05.06.2014, unique ID RFMEFI60714X0028), "Development of experimental combined tool designs using the super hard composite materials for the efficient rock destruction".

References

[1] Guericke B G, Krestovozdvizhensky P D, Dvornikov L T and Klishin V I 2015 *Depreciation and Lifetime of Cutters for Rock Destruction* Certificate of database state registration 621092 p 1

[2] Krestovozdvizhensky P D, Klishin V I, Nikitenko S M and Gerike P B 2014 *Physical and Technical Problems of Natural Resources Mining* **6** 107–115

[3] Anokhin A S, Strelnikova S S, Kukueva Ye V et al *Glass and Ceramics* **8** 26–29

[4] Dvornikov L T, Nikitenko S M, Korneyev V A and Korneyev P A 2015 *Drilling Bit with a Cutting Insert in the Form of Cassinian Elliptical Oval* patent appl. (Russian Federation) 2015125831 p 7

[5] Dvornikov L T, Nikitenko S M, Korneyev V A and Korneyev P A 2015 *Drilling Bit* patent appl. (Russian Federation) 2015125832 p 7

[6] Alimov O D and Dvornikov L T 1976 *Drilling Machines* (M.: Mashinostroenie) p 295

[7] Dvornikov L T, Nikitenko S M, Korneyev V A and Korneyev P A 2016 *Cutting Plate of a Drilling Bit* patent appl. (Russian Federation) 2016103526 p 7

[8] Dvornikov L T, Nikitenko S M, Korneyev V A and Korneyev P A 2016 *Cutting Plate of a Drilling Bit* patent appl. (Russian Federation) 2016103561 p 7

[9] Klishin V I, Dvornikov L T, Nikitenko S M and Korneyev V A 2015 *Evaluation of Rock Destruction Tool Equipped with Inserts of Superhard Composite Materials from the Standpoint of the Energy Consumption Process Mining Information-analytical Bulletin* pp 132–142

[10] Dvornikov L T and Naumkin N I 1983 *Method of L T Dvornikov – N I Naumkin for Determination of Rock destruction Energy Consumption and Device for its Realization* Invention certificate (Russian Federation) 1173826 p 6

[11] Dvornikov L T, Nikitenko S M, Korneyev V A and Korneyev P A 2015 *Device for Determination of Energy Consumption of Rock Distruction* patent (Russian Federation) 157386 p 7

[12] Dvornikov L T, Nikitenko S M, Korneyev V A and Korneyev P A 2015 *Device for Determination of Energy Consumption of Rock Distruction* patent (Russian Federation)

2587391 p 6

[13] Dvornikov L T, Nikitenko S M, Korneyev V A and Korneyev P A *Device for Determination of Energy Consumption of Rock Distruction* patent (Russian Federation) 157385 p 7

[14] Dvornikov L T, Nikitenko S M, Korneyev V A, Gusev M M and Korneyev P A 2016 *Energomer v 1.0* appl. for computer program registration (Russian Federation) 2016612577 p 8

Evaluation of elements loading in the metal structures of powered support units

M S Nikitenko[1,2]

[1] Federal Research Centre of Coal and Coal Chemistry of Siberian Branch of the Russian Academy of Sciences, 10 Leningradsky Ave, Kemerovo, Russia, 650065
[2] T.F. Gorbachev Kuzbass State Technical University, 28 Vesennyaya Street, Kemerovo, Russia, 650026

E-mail: schum24@rambler.ru

Abstract. In the paper the evaluation results of elements loading in the metal structures of powered support units are presented performed in the laboratory and plant conditions using the developed test samples of portable strain-gauge transducers on the basis of a spring element and a mobile multifunctional automated strain-gauge system.

1. Introduction

Performance of the units of powered roof support, operating elements of heading and shearing machines and other mining equipment to a large extent depends on the safety margin provided by design, manufacturing quality, nature and magnitude of operational loads, timely repair and restoration work.

Possibility of the rapid assessment of actual loads acting on the elements of metal structures both at the stage of manufacturing during the strength tests and during the operation is important for the improvement of existing equipment and development of new one.

Metal structures loading is characterized by a set of acting loading forces and can be assessed through deformations and stresses on the surfaces of their element and in the joints.

At present the existing method of strain-gauge control used during industrial tests does not allow the location of deformation measurements of the investigated metal structures surfaces to be immediately corrected due to impossibility of the repeated use of the gluing strain gages as the primary measuring transducers (PMT). Besides, the gage sensitivity factor k of the used wire strain gages made of constantan and nichrome is usually not higher than 2-2.5 units.

Thus, the question of direct assessment of the elements loading in the metal structures of powered support units is not fully resolved and has prospects for its development. One of the decisions, which allows the PMT mobility to be provided, i.e. possibility of its transfer and reinstallation, is to use the portable tensometric transducer (PTT) with an elastic element. However, inclusion of the elastic element into the measurement system greatly reduces the transducer sensitivity [1]. In contrast to the traditionally used strain gages, which are associated with the control object by every point, the sensitivity and metrological characteristics of PTT directly depend on the elastic element design, fixing method and the strain gauges used.

2. Development and methodology

In FRC CCC SB RAS laboratories of coal mechanical engineering, coal deposits geo-mechanics the development and experimental implementation of a technical complex used for diagnosis of the coal mines and open pit mines equipment [2]. By the present moment, the software and instrumental complex of a mobile multi-function automated strain gage system (MFATS) and PTT with an arch type of elastic element are brought to practical implementation in the form of prototypes.

As part of this work, in order to identify the geometric parameters of the arch elastic element, providing maximum sensitivity of the PTT, a study of influence of the elastic element configuration on deformation level in the strain gage installation zone was carried out. Sensitivity of the elastic element in the general form is characterized by the deformation level in the strain gages installation zone at similar deformations of the control object. As a result, the configuration of the arch PTT elastic element with a base of 100 mm with a concave arch of variable section and radial arches conjugation, as shown in Figure 1, is substantiated. Calculation of geometric parameters of elastic elements models was made by the finite element method.

Depending on the stated tasks of analysis two methods for fixing the elastic element on the control object surface are proposed and tested. The first one – the elastic element with the threaded connection is secured by special screw stands, which are reusable glued platforms with a threaded element. The second one – the elastic element is fixed to the surface with the permanent magnets made of the magnetic alloy NdFeB. It is more appropriate in cases of short-term static loads, such as the experimental determination of the of metal structures elements deformation.

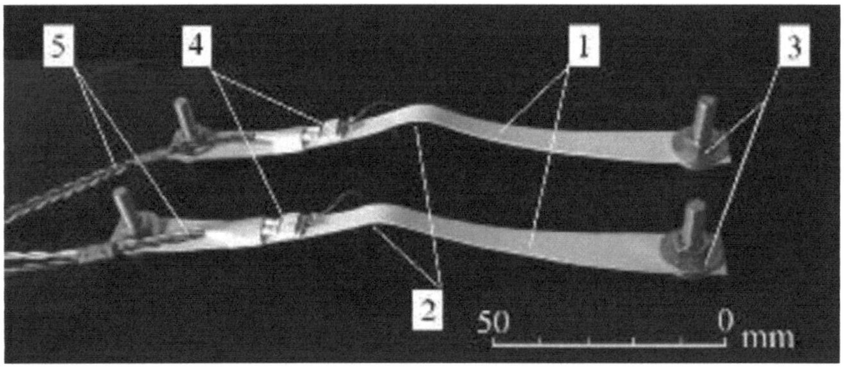

Figure 1. Test samples of arch PTT: 1 – elastic element, 2 – strain gage installation zone, 3 –fixing elements, (screw stands are shown), 4 – connecters, 5 – connecting wires.

The main characteristics of PKS type (Russia) and KSP (Japan) are studied in statics and dynamics in relation to the strain gages for PTT with the use of certified instruments and equipment of Siberian Aeronautical Research Institute n.a. S.A. Chaplygin [3] (installation for of the glued strain gages, certificate RU.E.28.007.A No. 27992, GRSI No. 34927-07 and tenso-system MMTS-64.01, certificate RU.C.34. 007. A No. 44412, reg. No. 21760-01) [4]. It is shown that semi-conductor KSP strain gages can be used to measure the structures relative deformations including short-time shocking processes.

As a result, on the basis of KSP strain gages and the proposed configuration of the arch elastic element PTT are manufactured on a screw stand and magnetic mount. Based on the

IOP Conf. Series: Earth and Environmental Science **45** (2016) 012007 doi:10.1088/1755-1315/45/1/012007

research of static characteristics of the manufactured PTT the calculation of tensor-sensitivity coefficient values k_i by the formula was made:

$$k_i = \left| \frac{\dfrac{(N_{maxi} - N_{0i})}{K_{coni}} + (R_{0i} - R_{ri})}{R_{ri} \cdot \varepsilon_{max}} \right|, \quad (1)$$

where N_{maxi} – value of the measuring tenso-system code for i-th PMT at maximal deformation; N_{0i} – value of the measuring tenso-system code for i-th PMT on the unloaded beam; K_{coni} – conversion factor for i-th channel of the measuring tenso-system; R_{0i} – resistance of i-th PMT with the unloaded beam; R_{ri} – rated resistance of i-th PMT; ε_{max} – maximal relative beam deformation.

For PTT k_i when mounted on a screw stand was 29.76, with pressure (magnet) mounting – 26.71, that in spite of elastic element influence greatly exceeds k value of standard PKS-type strain gages.

Next, operation of the test sample of mobile tensometric system and PTT was checked on the metal structure element of powered support unit in laboratory conditions and also in the process of assembling the powered support unit in the industrial conditions. MFATS prototype includes PTT, transmitter, ADC, computer software "Stend-Info" for registration, visualization, and analysis of experimental data (FIPS certificate No. 2011618442).

Testing method development of deformation rapid analysis by the output MFATS prototype signal U_{out} was performed with the help of a lateral bending of the metal structure weld sample at the laboratory loading bench (Figure 2a). As a sample the powered support unit MKYu.2U-12/25 traverse lever was used. Deformation evaluation of the lever front surface in twenty five control zones is presented. With this aim the researched front surface conventionally as a matrix was divided into control areas according to Figure 2b.

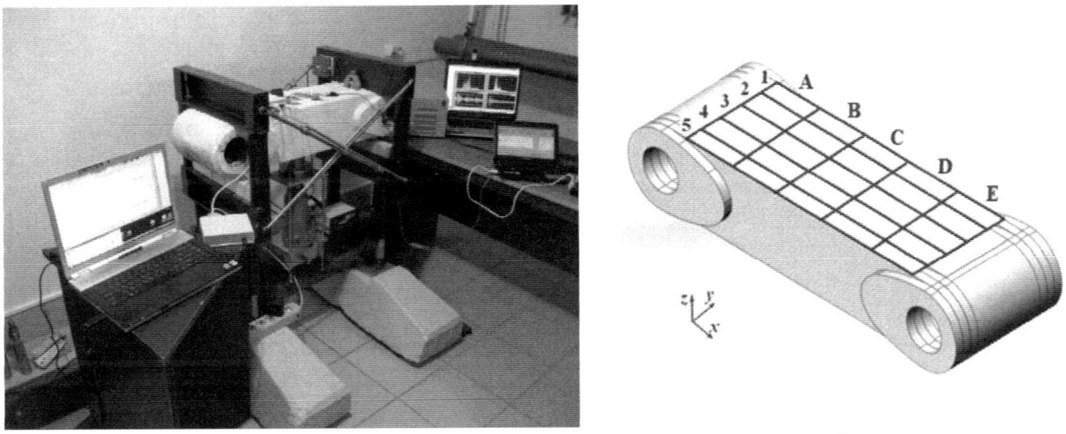

a b

Figure 2. Rapid analysis of the unit MKYu.2U-12/25 traverse lever in laboratory conditions: a – laboratory loading bench with MFATS prototype equipment, b – partitioning scheme of traverse lever front surface of the unit MKYu.2U-12/25 into control zones.

The idea of the method consists in fixing the PTT with the magnets at each control zone on metal structure surface and registration of output U_{out} signal which characterizes deformations along the axis coinciding with PTT axis when the sample is loaded.

Testing of the MFATS prototype at Yurginsky Machine Engineering Plant was carried out during the test of powered support unit MKYu.2U-07/19 at the "Gluckauf" bench for research and certification tests of powered support units in accordance with GOST R 52152-2003 (Figure 3).

With this purpose at the control points of the roof lower surface of the powered support parallel to the authorized tensometric system with the glued tenso-resisters of PKS type – 5 TPP – 1 and the equipment of mobile tensometric system were installed: transmitter – 2, ADC – 3, temperature indicator – 4, as it is shown in Figure 3b.

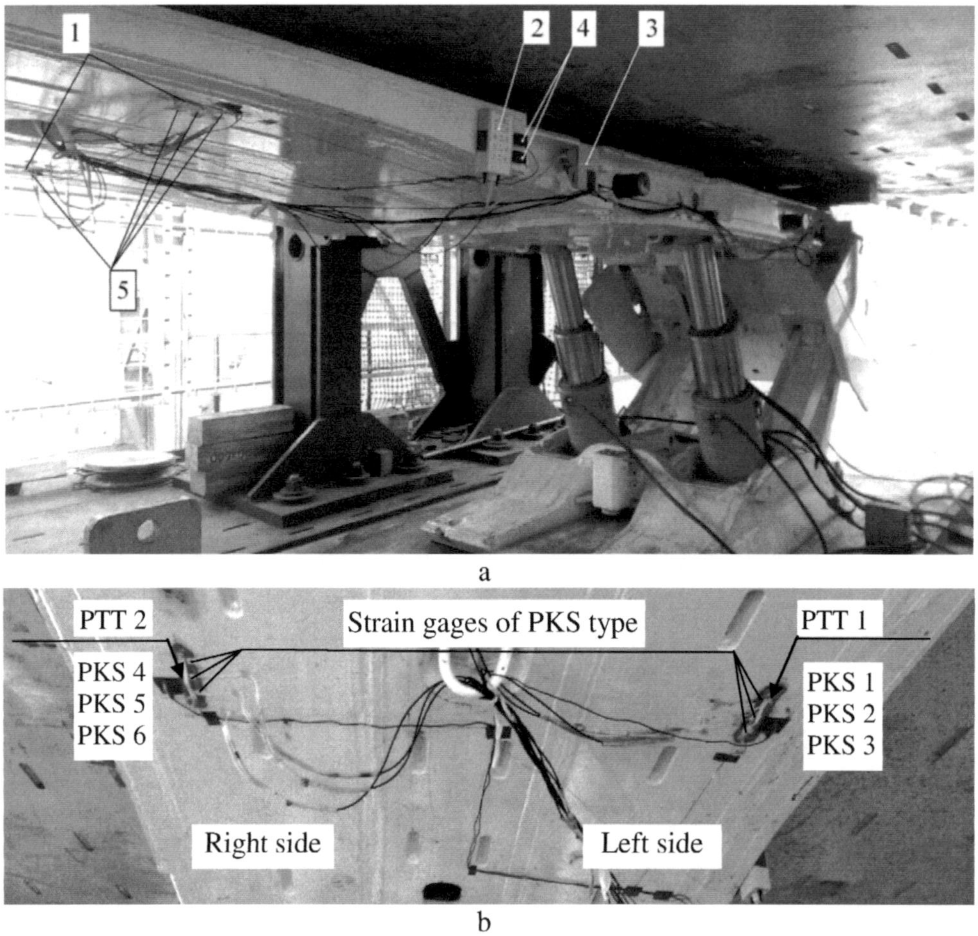

Figure 3. MKYu. 2U-07/19 unit at the loading test bench "Gluckauf" of LLC "Yurginsky Machine Engineering Plant": a – with the installed PTT and MFATS prototype equipment, b – PTT and strain gages (PKS type) layout in the control zones of the roof lower surface.

After this the support unit loading cycle was conducted at the bench for the application of conformity scale between the MFATS prototype – U_{out} signal and mechanical stresses values at the control points. The support unit material is 30HGSA steel. Such mechanical calibration was carried out on the basis of standard tensometric system data. As a result of the loading cycle the calibration chart of the MFATS prototype is built (Figure 4).

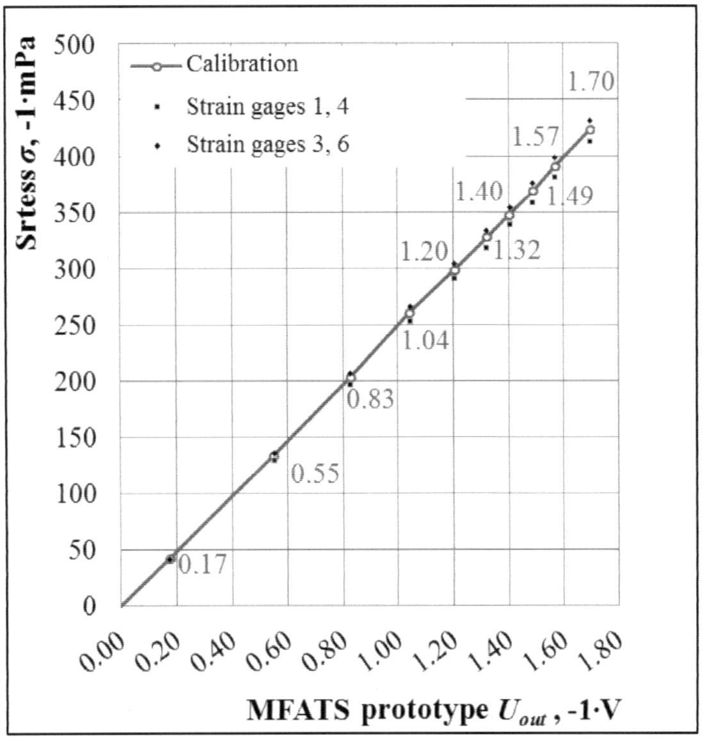

Figure 4. Calibration chart.

For the current level K_{coni} of the MFATS measuring transducer the dependence of mechanical stresses σ of the surface on U_{out} MFATS was expressed as:

$$\sigma = 1.149 - 249.1 \cdot U_{out}. \tag{2}$$

After that the support unit according to the methods approved by the plant was loaded in a stepwise manner until the pressure in the props became equal to 5, 17, 26, 34, 39, 43, 46, 49, 52 and 56 MPa with the intermediate unloading up to 5 MPa. In the process of sample testing MFATS prototype U_{out} in the control points depending on the load for each of the installed PTT.

3. Results and discussion

The data acquired as a result of the rapid analysis are summarized in the surface chart shown in Figure 5a, which allows the actual deformed state of the studied front surface to be visualized. These data allow the comparison of evaluation with the results of the modeling of this loading by the finite element method to be performed (Figure 5b).

According to the results of numerical error estimation of lever surface rapid analysis it is stated, that relative measurement error was less than 12 %.

On the side lever surface such rapid analysis allowed to determine not only the amount but also the type of deformation (expansion/contraction) for each control zone, depending on the longitudinal or transverse surface partition.

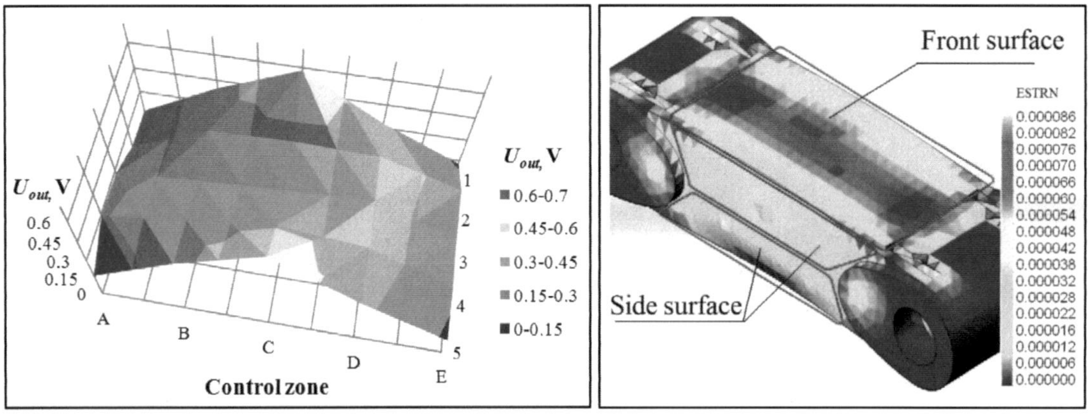

Figure 5. Rapid analysis method: a – the output stress U_{out} in the study of the lever front surface along the axis x under load, b – the calculated values of equivalent deformations of support unit lever when loaded on the bench by finite element method.

Test results obtained in the industrial conditions allowed the mechanical stresses dependence in the control points of the load for each of the installed PTT to be received. It was determined that maximum stress value according to the PTT data totaled 428 MPa (Figure 6).

In the comparison of the evaluation of the measurement results with the calculated values the average data error did not exceed 5.1 % for the PTT.

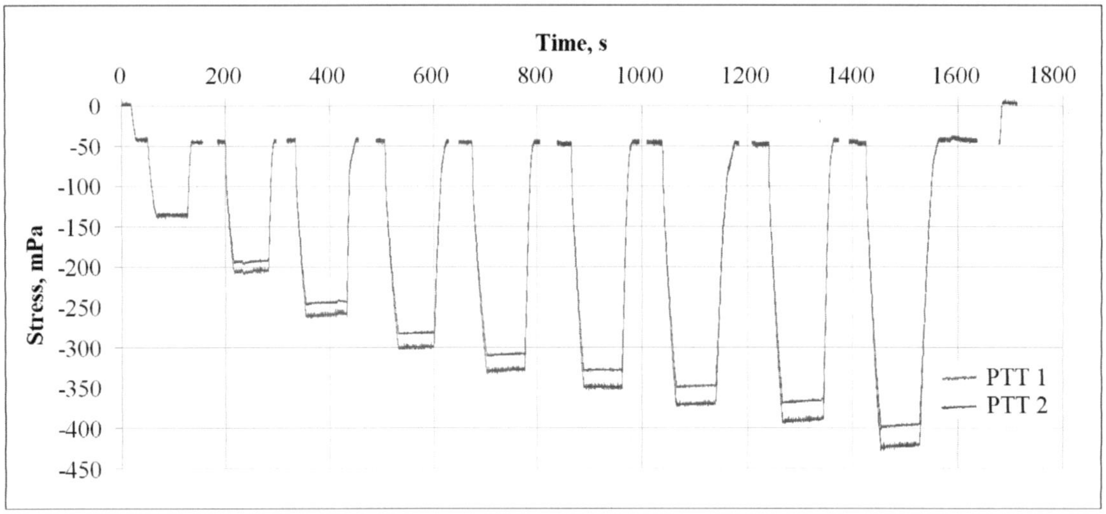

Figure 6. The character of stress changes, registered by the installed PTT on the left (PTT1) and right (PTT2) side of the roof section.

When comparing the obtained data with the results of the regular strain gauge system the errors for PTT were not more than 2 %. Wherein PTT may be repeatedly used and replaced along the studied surface in order to adjust the measuring area and for the rapid assessment of the controlled object loading.

4. Conclusions

The use of modern semiconducting materials and nonlinearity characteristics of compensation software algorithms allow highly sensitive PTT to be created, which help to study the elements loading in the mining machines metal structures under static loads at the bench. It can be used as an experimental supplement to the theoretical and practical evaluation and forecasting of the elements loading of powered support unit in the working face [5].

The proposed rapid analysis using the developed arched PTT allows the deformation along the axis to be assessed at any place of the studied element flat surface of metal structure, which coincides with the PTT axis with an error of no more than 12 %. It can be used for an operational definition of the greatest deformation zones of elements in the metal structures of mining equipment or for pre-visualization of the deformed state of control object.

The proposed prototype of a software-instrumental complex of mobile strain gauge system when placing PTT on the inner surfaces of roof planes, bases, arm traverses and powered support section guards provides measurement of mechanical stresses of metal structures the in the range from 0 to 450 MPa with an error of no more than 6 % from the calculated values.

5. Acknowledgements

Software-instrumental part of MFATS prototype is implemented through funding of the project "Development of the mobile microprocessor strain gauge system for control of the state of industrial structures" in the framework of a "START" program coordinated by the Fund for Assistance to Small Innovative Enterprises in Science and Technology (State Contract number 11089p/8345 from 01.10.2012).

The work was financially supported by the Ministry of Education and Science of the Russian Federation within the Federal Targeted program for Research and Development in Priority Areas of Development of the Russian Scientific and Technological Complex for 2014-2020 (agreement No. 14.607.21.0028 from 05.06.2014, unique ID RFMEFI60714X0028).

References

[1] Andreeva L E 1962 *Elastic Elements of Instruments* (Mashgiz, Moscow) p 236

[2] Nikitenko M S, Knyazkov K V, Ababkov N V and Ozhiganov E A 2013 *Mining Informational and Analytical Bulletin (Scientific and Technical Journal)* **S6** 447–456

[3] Sibnia.com. 2015 *Scientific Base* http://sibnia.com/scientific-base/airframe-fatigue-and-static-strength-research-division.php

[4] Stepanova L N, Lebedev E U, Kabanov S I, Kanifadin K V et al 2013 *Sensors & Systems* **1** (164) 28–33

[5] Sisoev N I and Turuk Yu V 2016 *Development and Construction of Powered Supports for Plough Complexes* (Novocherkassk: SRSPU(NPI)) p 226

IOP Conf. Series: Earth and Environmental Science **45** (2016) 012008 doi:10.1088/1755-1315/45/1/012008

Comparison of outburst danger criteria of coal seams for acoustic spectral and instrumental forecast methods

A V Shadrin and Yu A Bireva

Federal Research Centre of Coal and Coal Chemistry of Siberian Branch of the Russian Academy of Sciences, 10 Leningradsky Ave, Kemerovo, Russia, 650065

E-mail: ashadr1951@mail.ru

Abstract. Outburst danger criteria for the two methods of current coal seam outburst forecast are considered: instrumental – by the initial outgassing rate and chippings outlet during test boreholes drilling, and geo-physical – by relation of high frequency and low frequency components of noise caused by cutting tool of operating equipment probing the face area taking into consideration the outburst criteria correction based on methane concentration at the face area and the coal strength. The conclusion is made on "adjustment" possibility of acoustic spectral forecast method criterion amended by control of methane concentration at the coal face and the coal strength taken from the instrumental method forecast results.

1. Introduction

Safety rules of Russia's coal and shale mines prescribe to equip the mine openings with the systems and means providing the safe mining operations. These systems must be combined into multifunctional safety systems (MFSS) which are to perform among other operations the gas-dynamic phenomena (GDP) [1] control and forecast.

One of the most dangerous GDP types is sudden outbursts of coal and gas. The current outburst forecast can be fulfilled by two groups of methods:

- instrumental, based on the face area control with a test borehole, and
- geo-physical which evaluates the condition of the face area by probing it with acoustic or electromagnetic waves.

The widespread introduction of geophysical forecasting methods is constrained by the difficulty of substantiation of geophysical outburst criterion for specific geological and mining conditions.

The aim of the present work is to show that the most comprehensive technological solution will be to "adjust" the outburst danger criterion of geophysical acoustic spectral method, amended with the gas factor and coal strength control, according to the results of instrumental method, the initial outgassing rate and the chippings outlet during test borehole drilling.

2. Outburst danger criteria of instrumental and acoustic spectral methods of outburst danger forecast

When developing methods to predict coal and gas sudden outbursts, their signs determined during multi-year study of their manifestation, are taken into account. They include [2]:

- excessive pressure on the mine opening support;
- shocks, crackles in the rock mass;
- coal face peeling, squeezing out, coal pieces peeling and pouring out from the coal face;

- jamming of the drilling tool, its pushing out or pulling inside the borehole;
- coal strength reduction;
- increase of gas inflow while drilling or coal cutting works.

Outburst signs analyses proves that the main factors determining the outburst danger are stressed condition, inter-bedding gas pressure and the coal strength. To control these parameters at Russian mines direct (or instrumental) and indirect (geo-physical) methods are used.

Among instrumental methods the most reliable forecast at the mines in eastern Russia is provided by the method based on the interval measurement (in 1 m) of the initial outgassing rate and the chippings outlet as a result of control holes drilling [3]. Wherein the initial outgassing rate characterizes gas factor of the outburst danger and the chippings outlet – the stressed condition of the seam and coal strength. The outburst danger criterion at current forecast using this method at Kuzbass mines, which is based on a long-term processing of experimental data, is described by the following equation [3]:

$$R_1 = (S_{max}-1.8)(i_{max}-4) - 6 = 0, \quad (1)$$

where S_{max} is the maximum value for the chippings outlet, l/m; i_{max} – maximum value for initial outgassing rate, l/min·m. This dependence graph is shown in Figure 1.

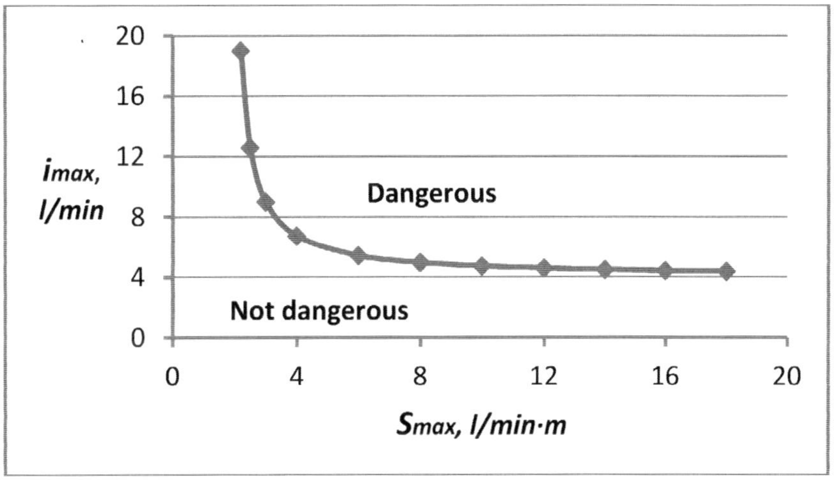

Figure 1. Dependency graph of outburst danger index $R_1 = (S_{max}-1.8)(i_{max}-4)-6=0$.

The main drawbacks of this method are periodical character of the forecast, long duration of the procedure during which the coal face operations need to be stopped.

The geo-physical methods are more advanced as they do not interfere with the mining operations. At present time three such main methods are known: the method of acoustic emission (AE) controlling the crack development process; gas analyzing method – by methane concentration in the mine atmosphere measured with air and gas control (AGC) instruments which control the factor of gas outburst danger; and acoustic spectral – by relation of high-frequency and low-frequency noise amplitudes of the operating mining equipment, which controls the stressed condition. The essence of the advantages and disadvantages of these methods are described in [4]. Their main disadvantage is the problem of justification of the geophysical outburst criterion for specific geological and mining conditions.

The simplest way to justify this criterion is to collect the experimental data while heading the galleries through the seam sections for which with the help of other methods the category "dangerous" and "not dangerous" was assigned. However, this approach seems to us time consuming and, therefore, not technological. A much more advanced is the method which quantitatively establishes the degree of outburst danger for the exact section by a reliable instrumental method and then according to this quantitative danger degree the outburst danger forecast of geo-physical criterion is "adjusted". To implement this approach we selected a acoustic spectral method which showed a high accuracy of the forecast compared with other geophysical methods at the coal mines in Donbass [5] and is in the process of implementation at some mines of JSC "SUEK-Kuzbass" in Kuzbass.

However, in this classic manner of application (by amplitudes relation of high- and low-frequency spectrum parts of operating equipment noise) this method only qualitatively controls the seam stressed state (at present there is no method for quantitative determination of outburst danger criterion). Therefore, we attempted to develop a correction procedure for the outburst danger index of a acoustic spectral method with regard to the gas ratio and coal strength, estimated by instrumental techniques of forecasting.

Outburst indicator of the acoustic spectral method was determined for the two main processes of outburst situation preparation: development of cracks at the face area and creation of coal block structure (the first process) and extrusion of coal blocks (lamps) into the working in the mouth of the future outburst cavity (the second process) in the following form:

$$K_c = \frac{A_h}{A_l} = e^{-C\frac{\sigma_l}{\sigma_c}d}, \qquad (2)$$

where K_c – the current value of the outburst danger index; A_h A_l – amplitudes of acoustic noises, produced by the operating mining equipment, measured respectively at high and low frequencies; σ_l and σ_c respectively, the average limit and the current stresses;

$$C = \frac{\alpha_0 \beta (f_h - f_l)}{f_0}, \text{ m}^{-1}; \qquad (3)$$

f_h and f_l – the characteristic frequencies from the ranges of upper and lower working frequencies of the acoustic noise source, Hz; α_0 – decay coefficient at a certain frequency f_0 which belongs to a range of registered frequencies, m^{-1}; β – the proportionality factor determined by the rock mass properties; d – distance between the noise source affecting the operating face (combine machine, coal plough, pick hammer, drill bit) and the receiver installed in the wall of the working at a certain distance from the face, m.

The obtained approximate expressions for the criteria determination of these processes are given in [4]. To improve the accuracy of determining the outburst criteria for acoustic spectral method the studies were carried out, the main results of which are as follows.

As a physical criterion of the outburst for the first process – a crack development at a distance x_{cr} from the working face, due to external compressive loads $\sigma_1(x)$ - $\sigma_3(x)$ (a plane problem) and internal gas pressure P, – the criterion, proposed by I.M. Petukhov and A.M. Linkov, was taken [6]:

$$\psi_{cr}\frac{P - |\sigma_3|}{\sigma_s} + \frac{|\sigma_1|}{\sigma_p} = 1, \qquad (4)$$

where $\psi_{cr}(x_{cr}/l) = k_1/k_1^{\infty}$; k_1 and k_1^{∞} are stress intensity factors at the observation point and outside the working zone influence; x_{cr} – critical distance from the working face, in the rock face of which the outburst develops (in this case a crack starts developing); l – the crack half-length; σ_p and σ_s are strength limits for compression and extension; $\sigma_1(x)$ и $\sigma_3(x)$ – main stresses.

The following expression was obtained for this criterion to determine the average stress limit σ_l depending on coal strength and in-situ gas pressure which cause the crack growth:

$$|\sigma_l| \simeq L \left| \left[\frac{0.3}{\psi_{cr}}\left(\frac{q}{110-q}\right) - P\right] \right|, \text{MPa} \quad (5)$$

where q is coal strength, determined with strength meter designed at A.A. Skochinsky Institute of Mining, $L = \frac{2+\lambda}{3\lambda}$ – the average coefficient of lateral thrust, λ – lateral thrust coefficient.

Using the dependence of gas flow from the face surface on the in- situ gas pressure proposed by V.V. Khodot [7] the expression was obtained relating the in-situ gas pressure P to the methane concentration Ω in the working atmosphere:

$$P = D\sqrt{\frac{Q\Omega}{\xi_i}}, \quad (6)$$

where: Q – local ventilation fan airflow, ventilating the working, m^3/s; ξ_i – coefficient taking into account the impact degree (proportion of fresh exposed face area) of i- type equipment (combine, plough, pick hammer, drilling bit diameter, etc.) on the face, $0 < \xi_i < 1$; Ω – the current value of methane concentration measured with a methane meter, %; $D = m\sqrt{k_0\eta P_{at}} \cdot x_{cr} \exp(-\varphi x_{cr}) / \sqrt{100 S_f}$, $Pa \cdot s^{1/2} m^{-3/2}$.

Substituting (5) and (6) in (2), we acquired the following expression for determining the current limit of outburst danger indicator (the initial stage of cracks development):

$$K_{l,c,1} = e^{-cd\left[\frac{1}{0.1\left(\frac{q}{110-q}\right) - \frac{\Psi_{cr}}{3} D\sqrt{\frac{Q\Omega}{\xi_i}}}\right]}. \quad (7)$$

Estimation of parameter D can be derived from the following considerations. The coal seams, dangerous due to their manifestation of the gas-dynamic phenomena could be related to the following three groups: dangerous because of rock bursts manifestation, dangerous because of sudden outbursts and rock bursts and because of sudden outbursts. The first group is characterized by high coal strength and high existing stresses, the gas pressure can vary within very wide limits. The second group on the contrary is characterized by low coal strength, high gas pressure with a rather high value of average stress. The third group is characterized by the mean values of coal strength with significant (sufficient) medium stresses and gas pressure. For the third group of coal seams, which are most typical for Kuzbass, we can suggest an approximately equal influence of rock pressure forces and in-situ gas pressure on the outburst danger index. Therefore, from the equality of common denominator members in the exponent (7), the parameter D is easily determined. We introduce the notation:

$$\text{æ} = 1/q. \quad (8)$$

Then from (7) we obtain an expression for determining a family of curves for different values $K_{l,c,1}$, connecting methane concentration with the reciprocal value of coal strength:

$$\Omega = \frac{9\xi_i}{\Psi_{cr}^2}\left[0.01\left(\frac{1}{110\text{æ}-1}\right)^2 + 0.2\left(\frac{1}{110\text{æ}-1}\right)\frac{cd}{lnK_{l,c,1}} + \frac{c^2d^2}{ln^2 K_{l,c,1}}\right]. \quad (9)$$

The graph of the curves family in form (9) is shown in Figure 2.

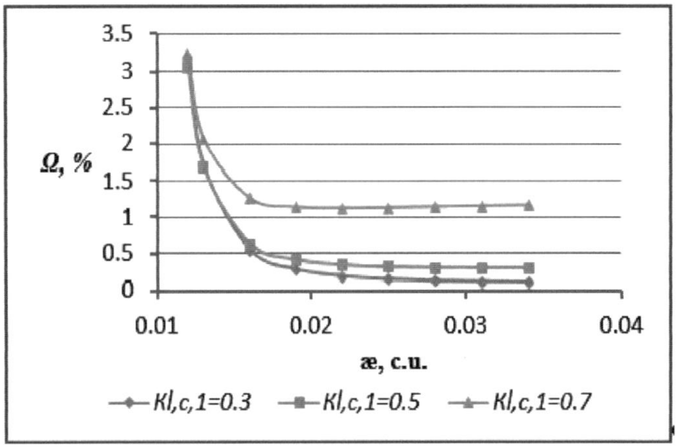

Figure 2. Family of curves Ω=Ω(æ) for values $K_{l,c,1}$=0.3; 0.5 and 0.7 in the process of crack development with the following parameters: ξ_i=0.1; Ψ_{cr}=5; Q= 10 м^3/s; α_0=1.3 m^1; β=0.07; D=0.01 MPa·s$^{1/2}$·m$^{-3/2}$; d=10 m; q=30-100 c.u.

Comparing the curves in Figure 1 and Figure 2, described respectively by expressions (1) and (9), we can see that the geo-physical outburst danger criterion of acoustic spectral method determined for the crack development process has the same character of dependence on gas factor and coal strength as well as the criterion of instrumental method by the initial outgassing ratio chippings outlet during test boreholes drilling. The main reason of differences is conditioned by the fact that in the function i_{max}=f(S_{max}) S_{max} parameter depends on the stress state and coal strength, whereas in function Ω = f(æ) æ parameter depends only on the strength of coal.

Similar studies have been conducted to establish the outburst danger criterion of acoustic spectral forecast method for the second preparation process of the pre-outburst state – coal blocks (lamps) extrusion into the working from the face surface in the mouth of the future outburst cavity. The following relationship was taken as the physical criterion of this process [4]:

$$K_{cr} = \frac{F_a}{F_p} = \frac{F_1+F_2 \pm |F_3|}{F_4} \qquad (10)$$

In this criterion the active (triggering) in the numerator and passive (preventing) forces in the denominator forces are shown that press out the coal block into the working from the face. The first summand in the numerator of F_1 corresponds to the force of lateral thrust (horizontal component of normal stresses), the second summand F_2 – to the gas pressure force, the third F_3 – to the gravity of the pressed out coal block, and denominator F_4 – describes cohesion force and coal internal friction in the massif. To simplify the task it was assumed that the pressed out coal block has a cylindrical shape with the effective radius r_e and thickness x_1. The sign in front of F_3 depends on the working type – downward (-) or upward (+).

For this criterion, the following expression was obtained to determine the average stress limit σ_l, depending on the coal strength and in-situ gas pressure at which the coal block is pressed out into the working:

$$\sigma_l \lesssim \frac{L}{cos\delta}\left[9{,}5\left(\frac{q}{110-q}\right)10^6\frac{x_1}{r_e} - P \mp \gamma_c x_1 sin\delta\right], \qquad (11)$$

where δ – the angle of the opening axis to the horizon ; γ_c – coal specific weight.

Substituting (6) and (11) into (2), we received the following expression for determining the current limit value of the outburst indicator (the beginning of coal blocks pressing out):

$$K_{l,c,2} = e^{-C\left[\frac{10^8 \frac{x_1}{r_э} \mp \gamma_y x_1 \sin\delta}{10^7\left(\frac{q}{110-q}\right)\frac{x_1}{r_e} - D\sqrt{\frac{Q\Omega}{\xi_i}} \mp \gamma_y x_1 \sin\delta}\right]d} \; . \qquad (12)$$

We find parameter D the same way as it was done for the criterion of beginning of crack development, but assume that in this process the gas factor affects the outburst approximately 3 times stronger than the factor of stress state.

From (12) for horizontal mine workings (δ=0) by simple transformations we obtain the expression for the family of curves for different values of $K_{l.c.2}$, connecting the methane concentration with the reciprocal value of coal strength:

$$\Omega = \frac{10^{16}\ \xi_i x_1^2}{D^2 Q r_e^2}\left[\frac{c^2 d^2}{ln^2 K_{l,c,2}} + 0.2\left(\frac{1}{110æ-1}\right)\frac{cd}{lnK_{l,c,2}} + 0.01\left(\frac{1}{110æ-1}\right)^2\right]. \quad (13)$$

The graph of the curves family of form (13) is shown in Figure 3.

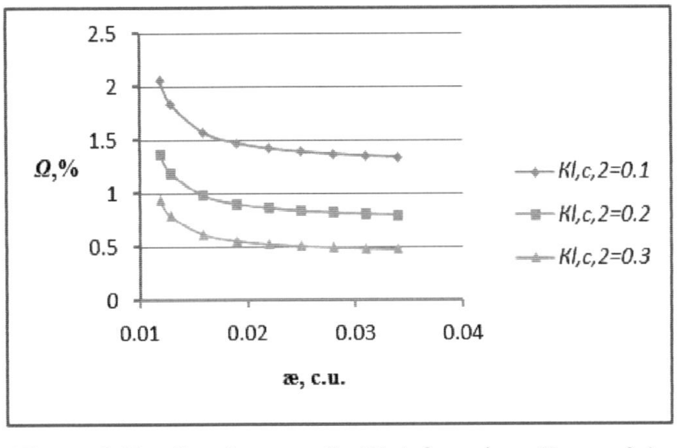

Figure 3. Family of curves Ω=Ω(æ) for values $K_{l.c.2}$ = 0.1; 0.2 and 0.3 in the process of a coal block pressing out with the following parameters: ξ_i=0.1; x_1=0.05 m; r_e=0,5 м; Q= 10 m^3/s; α_0=1.3 m^{-1}; β=0.07; D=1.0 MPa·s$^{1/2}$m$^{-3/2}$; d=5 m; q=30-100 c.u.

Comparing the curves in Figure 1 and Figure 3, described respectively by expressions (1) and (13), we can see that geophysical criterion of outburst danger of acoustic spectral method, established for the coal block pressed out into the working from the face surface in the mouth of the future outburst cavity, has about the same character of dependence on gas factor and coal strength as the criterion of instrumental method for the initial outgassing rate and chippings outlet during drilling the test boreholes. The differences, obviously, are due to the same cause as the differences in the dependencies in Figure 1 and Figure 2.

3. Conclusions

The obtained results prove the following. Firstly, outburst danger criteria of the acoustic spectral method for two considered processes of preparation of outburst dangerous situation, although have the same character of dependence on the basic influencing factors, differ quantitatively.

Secondly, the character of dependence of outburst danger factor on the influencing factors for the instrumental technique of the forecast as well as for geophysical one is much the same and allows the range of influencing parameter values to be divided into two zones: outburst dangerous and not dangerous.

Thirdly, the compared instrumental and geophysical criteria have quantitative outburst danger estimation which allows the criteria to be compared and, as we think, it will make possible to "adjust" geophysical criterion in a specific working in accordance with the criterion readings of the instrumental forecast method.

The obtained results will be used for development of methods for determining the critical index of outburst danger by acoustic spectral method for specific conditions of mine workings.

As acoustic spectral- method controls the stressed state of the rock mass by the noise spectrum of a working combine, installing noise receivers in ventilation and haulage galleries at a small distance from the face makes it convenient to control the dynamics of stressed state in the face area while controlling roof using such method as a directed hydraulic fracturing among others.

4. Acknowledgements
The paper was prepared with the financial support of the Ministry of Education and Science of the Russian Federation within the Federal target program "Research and development of Russia's scientific-technological complex of priority directions for 2014-2020", the theme is "Development of experimental combined tool constructions using superhard composite materials for efficient crush of mine rocks" (Agreement No. 14.607.21.0028 dated 05.06.2014, unique ID RFMEFI60714X0028).

References
[1] *Federal Regulations in the Field of Industrial Safety "Safety Rules in Coal Mines"*. Approved by Rostekhnadzor Order from 19.11.2013 No. 550 (Ekaterinburg: ID "Ural Yur Izdat") p 148
[2] *Prevention of Gas-dynamic Phenomena in Coal Mines: A Collection of Documents* 2011 Series 05. Issue 2 (M.: ZAO "Scientific and Technical Research Centre of Industrial Safety Problems") p 304
[3] Chernov O I and Puzyrev V N 1979 *Coal and Gas Outburst Forecast* (M.: Nedra) p 296
[4] Shadrin A V and Bireva Yu A 2015 *Science Intensive Development Technology and Utilization of Mineral Resources: Scientific Articles Collection* ed V N Fryanov (Novokuznetsk: SibSIU) pp 105–110.
[5] Mirer S V, Khmara O I and Shadrin A V 1999 *Acoustic Spectral Forecast of Coal Seams with Outburst Danger* (Kemerovo: Kuzbassvuzizdat) p 92
[6] Petukhov I M and Linkov A M 1983 *Mine Shocks and Outbursts Mechanics* (M.: Nedra) p 280
[7] Khodot V V 1961 *Sudden Outbursts of Coal and Gas* (M.: State Publishing House of Scientific-technical Literature for Mining) p 364

Construction of the approximant of complete diagram for rock deformation

A B Tsvetkov, L D Pavlova and V N Fryanov

Siberian State Industrial University, 42 Kirova Street, Novkuznetsk, Russia, 654007

E-mail: atsvet@mail.ru

Abstract. The function for approximation of the diagram of rock deformation, intended for nonlinear model of stress-strain state of coal geomassive, is developed, which is realized in the package of problem-oriented software. The computing experiments and the analysis of outcomes of numerical modeling are performed, confirming the correspondence of the calculated results with the measured settlings of underworked seams.

1. Introduction

Intensive technogenic influence on the geomassive leads to formation in the periphery areas of coal seam and in the zones of increase pressure of such geodynamic phenomena, as mine bumps, sudden outbursts of coal, rock and gas, caving and dynamic settlings of roof rocks with release from the goaf of dangerous gases in the form of shock air waves. The pointed phenomena limit the efficiency of the mechanized stopes equipped with expensive machines. As coal seam excavation leads to formation of discharge and bearing pressure zones around goaf, in which the rocks possess various resistance to extending and to compressive forces, then the consideration of a multimodule rock properties, the deformation law of which is defined by nonlinear equations, will allow qualitatively new information to be obtained about geomechanical state in the field of mine workings influence.

To solve this problem authors developed a nonlinear mathematical model of stress-strain state of geomassive [1]. For its defining correlations, constructed by the variable parameter method, it is necessary to elaborate function approximating a complete diagram of deformation.

The aim of the present work is to develop the function, approximating the complete diagram of rock deformation, which is constructed by the basic points, separating fields of elastic, elastoplastic, limiting and out-of-limit deformation.

2. Construction of approximants and computing experiments

The physical equations of the problem of plasticity theory in the process of linearization can be written in the form of formulas [2]:

$$\varepsilon_x = \frac{1}{E^*}\left[\sigma_x - \nu^*\left(\sigma_y + \sigma_z\right)\right]; \quad \gamma_{xy} = \frac{1}{\mu^*}\tau_{xy};$$

$$\varepsilon_y = \frac{1}{E^*}\left[\sigma_y - \nu^*\left(\sigma_x + \sigma_z\right)\right]; \quad \gamma_{yz} = \frac{1}{\mu^*}\tau_{yz}; \qquad (1)$$

$$\varepsilon_z = \frac{1}{E^*}\left[\sigma_z - \nu^*\left(\sigma_x + \sigma_y\right)\right]; \quad \gamma_{zx} = \frac{1}{\mu^*}\tau_{zx},$$

which provide recalculation of variable parameters according to dependencies:

$$E^* = \frac{3E}{2E\psi + 1 - 2\nu}; \quad \nu^* = \frac{\frac{1}{2} - \frac{1-2\nu}{E}\frac{1}{2\psi}}{1 + \frac{1-2\nu}{E}\frac{1}{2\psi}}; \quad \mu^* = \frac{1}{2\psi}, \tag{2}$$

where (ε_x, ε_y, ε_z, γ_{xy}, γ_{yz}, γ_{zx}) – components of a deformation tensor; (σ_x, σ_y, σ_z, τ_{xy}, τ_{yz}, τ_{zx}) – components of a stresses tensor; E – modulus of elasticity; ν – Poisson's ratio; E^*, ν^*, μ^* – variable parameters. $\psi = \frac{3\varepsilon}{2\sigma}$ – dependence of the generalized stresses σ on the generalized deformations ε; $\sigma=\sigma(\varepsilon)$ – function, defining the complete diagram of rock deformation.

On the basis of research in [3], the typical complete diagram of "stress-strain" has the form as of a diagram in Figure 1. On the complete diagram of deformation the characteristic points are marked: (ε_e, σ_e), (ε_s, σ_s), (ε_o, σ_o), where σ_e – the stresses corresponding to elastic limit; ε_e – limiting elastic deformations; σ_s – ultimate rock strength at contraction or extension; ε_s – limiting contraction deformations; σ_o – residual stresses; ε_o – residual deformations.

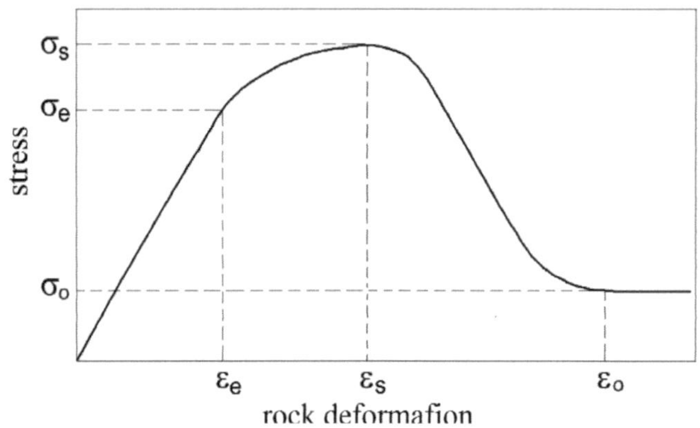

Figure 1. The complete diagram of "stress-strain".

For the analytical representation of diagrams of rock deformation the dependence in the form of rocks association in the form piecewise function is offered:

$$\sigma(\varepsilon) = \begin{cases} \sigma_e(\varepsilon), 0 \le \varepsilon < \varepsilon_e; \\ \sigma_s(\varepsilon), \varepsilon_e \le \varepsilon < \varepsilon_s; \\ \sigma_m(\varepsilon), \varepsilon_s \le \varepsilon < \varepsilon_o; \\ \sigma_o, \varepsilon \ge \varepsilon_o, \end{cases} \tag{3}$$

where σ – generalized stresses; ε – generalized deformations.

For stresses approximation $\sigma(\varepsilon)$ on each section the functions $\sigma_s(\varepsilon)$, $\sigma_z(\varepsilon)$ are elaborated. In the field of elastic deformation the connection between stresses and deformations is defined by the Hooke's law

$$\sigma_e(\varepsilon) = E \cdot \varepsilon, \tag{4}$$

where E – modulus of elasticity.

The elastoplastic section of the deformation diagram is defined by the parabola, the coefficients of which are calculated from the conditions of conformity with a straight line (4) in a point (ε_e, σ_e) of magnitudes of stresses and the first order derivatives

$$\sigma_s(\varepsilon) = \sigma_e + (\sigma_s - \sigma_e)\frac{\varepsilon - \varepsilon_e}{\varepsilon_s - \varepsilon_e} - \frac{(E + E^*)}{2(\varepsilon_s - \varepsilon_e)} \cdot (\varepsilon - \varepsilon_e)(\varepsilon - \varepsilon_s), \tag{5}$$

where E^* – deformation modulus, matching the limiting and out-of-limit diagram sections of rock deformation, which is defined by the dependence

$$E^* = E - \frac{2(\sigma_s - \sigma_e)}{(\varepsilon_s - \varepsilon_e)}. \tag{6}$$

Taking into account formula (6) the function $\sigma_s(\varepsilon)$ will be

$$\sigma_s(\varepsilon) = \sigma_e + (\sigma_s - \sigma_e)\frac{\varepsilon - \varepsilon_e}{\varepsilon_s - \varepsilon_e} + \frac{(\sigma_s - \sigma_e) - E(\varepsilon_s - \varepsilon_e)}{(\varepsilon_s - \varepsilon_e)^2}(\varepsilon - \varepsilon_e)(\varepsilon - \varepsilon_s). \tag{7}$$

The section of out-of-limit deformation is approximated by straight lines and matched with the parabola in the point (ε_s, σ_s):

$$\sigma_m(\varepsilon) = \sigma_o + (\sigma_s(\varepsilon_s) - \sigma_o) \cdot \frac{\varepsilon - \varepsilon_o}{\varepsilon_s - \varepsilon_o},$$

$$\sigma(\varepsilon) = \sigma_0. \tag{8}$$

The experimental diagrams of rock deformation are given in [4]. For evaluation of the results reconciliation with the in-situ data, the graphs were compared, and the correlation dependence between them estimated.

In figures 2a, b the graphs of experimental diagrams were compared for sandstones presented in in [4] and approximations of their piecewise functions constructed with the use of dependences (3) – (8).

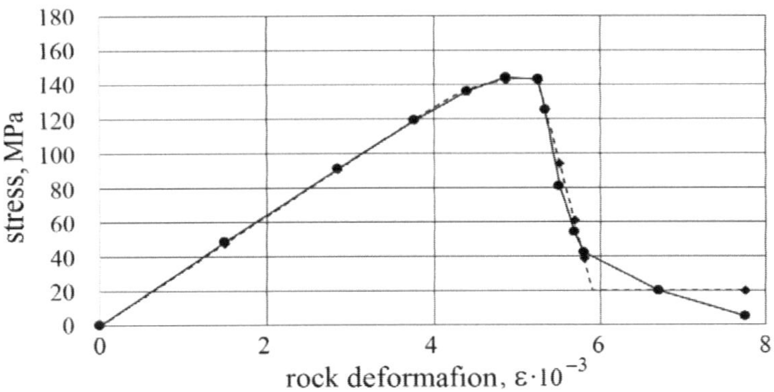

a

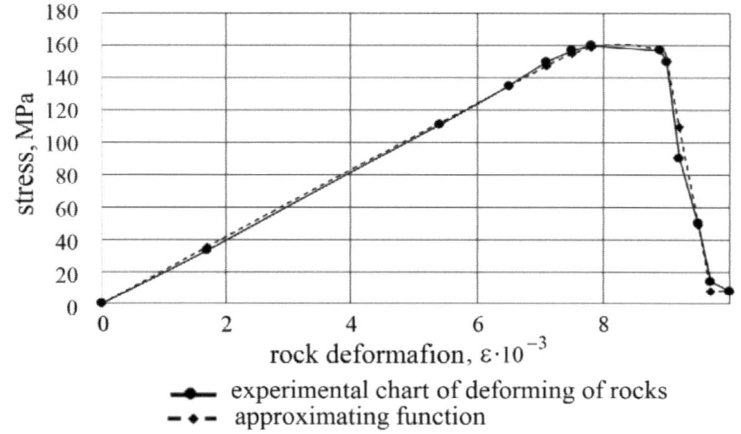

b

Figure 2. Approximation of experimental diagrams of sandstones deformation.

From the analysis of the graphs in Figures 2a, b the reconciliation of the results with the in-situ data is observed. The correlation factor is 0.994 and 0.996, respectively.

In Figure 3 the deformation diagrams for coal stretching and coal compression are given, constructed by the characteristic points (ε_e, σ_e), (ε_s, σ_s), (ε_o, σ_o) with the use of formulas (3) – (8).

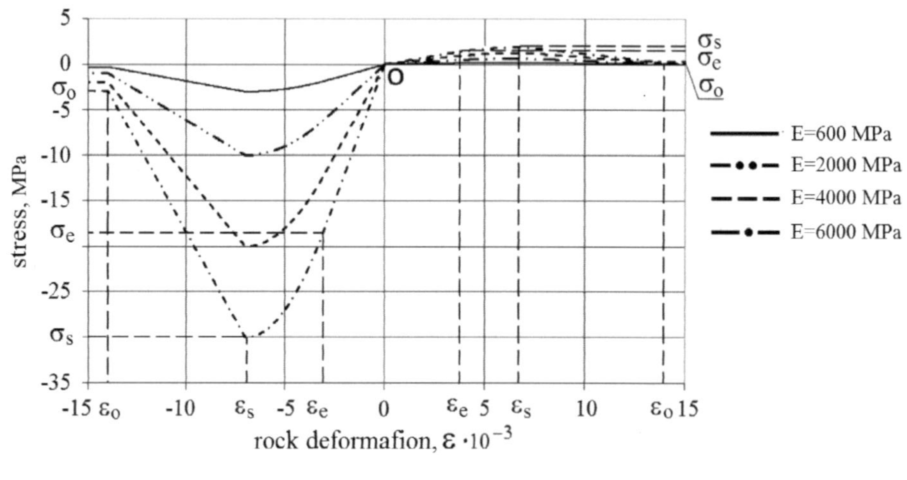

Figure 3. Deformation diagram of coal extension and contraction.

Approximant (3) is used in nonlinear mathematical model [1] of deformation of rocks in the process of their transition from elastic into elastoplastic, limiting and out-of-limit states.

To carry out calculation experiments the complex of the problem-oriented software was developed [6] which allow the numerical solutions of nonlinear problems of geomechanics to be obtained for evaluation of the stress-strain state of geomassive being under the technogenic influence and with the help of the developed modules to carry out all stages of the computing experiment.

To define quantitative reconciliation of numerical outcomes with the in-situ data computing experiments were performed, for which the mining and geological and technical conditions of seam 22 at the mine "Yubileynaya" in Kuzbass were accepted. The outcomes of numerical solution were compared with the in-sutu data of vertical movements recorded during advancement of the mechanized face from the installation chamber in the direction of the coal massive by rappers of the observant station R1–R7, established in the well at different depths (Figure 4).

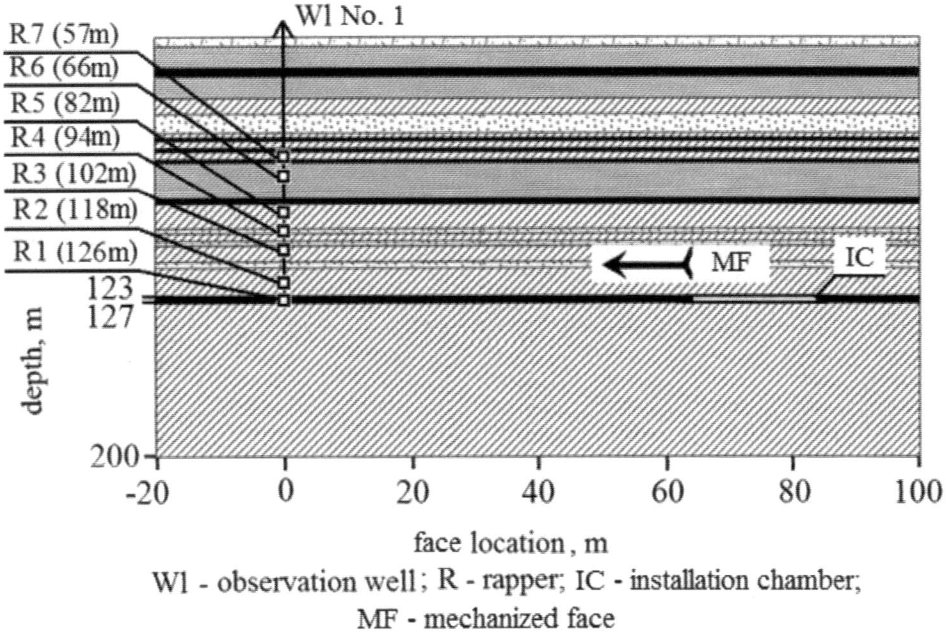

Figure 4. Scheme of a reppers location in front of the advancing face, seam 22, mine " Yubileynaya", Kuzbass .

In Figure 5 the graphs of vertical movements are shown, constructed according to the in-situ data (V^*(R1)–V^*(R7)) and the results of computing experiments (V(R1)–V(R7)).

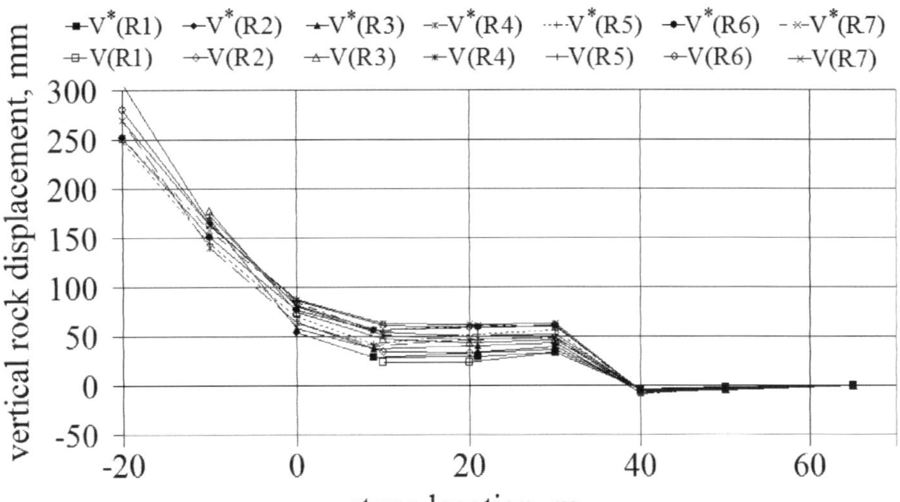

Figure 5. The vertical movements of rappers received by the data of mine measurements (V^*) and the results of computing experiments (V).

IOP Conf. Series: Earth and Environmental Science **45** (2016) 012009 doi:10.1088/1755-1315/45/1/012009

3. Results

As a quantitative measure of reconciliation of numerical modelling results with the in-situ data it is offered to use integral criterion of the form

$$I = \frac{\sqrt{\int_0^L (f^{solve}(x) - f^{nat}(x))^2 dx}}{\sqrt{\int_0^L f^{nat}(x)^2 dx}}, \qquad (9)$$

where L – length of the goaf; $f^{nat}(x)$ – the function approximating the in-situ data; $f^{solve}(x)$ – the function approximating the results of numerical modelling.

For evaluation of the reconciliation of numerical results with in-situ data by criterion (9) the approximant of the following form are constructed:

$$f^{nat}(x) = \sum_{i=1}^{N} a_i^{nat}|x - x_i|;$$

$$f^{solve}(x) = \begin{cases} \sum_{i=1}^{N} a_i^{e}|x - x_i|, & \textit{for solution of a problem of the elastic theory} \\ \sum_{i=1}^{N} a_i^{nlin}|x - x_i|, & \textit{for nonlinear solution,} \end{cases} \qquad (10)$$

where N – number of observations; a_i^{nat} – coefficients defined from the solution of equations system, constructed with use of mine measurements; a_i^{nlin}, a_i^{e} – coefficients defined for solutions of a nonlinear problem and a boundary value problem of the elasticity theory.

According to integral criterion (9) with the use of the constructed dependencies (10) the maximum relative error of numerical modeling is calculated, which for the results of the computing experiments, received by means of the elaborated nonlinear model, amounted 0.21, and for the known model constructed by the the defining proportions of the elastic theory – 0.90. The carried out analysis of results of modelling shows the adequacy of the constructed nonlinear model to the research subject.

4. Conclusions

The function, approximating the complete diagram of rock deformation, is developed. By its means in the nonlinear model of stress-strain state of coal-bearing massive [1], constructed by the method of variable parameters, the value of the intersecting modulus of deformation in equations (1) are algorithmically defined. Application of nonlinear mathematical model of a geomechanical state of the coal massive, including the elaborated piecewise function and the complex of the problem-oriented software developed on its ground, ensure results corresponding to the real data and allows the predictive estimates of stress-strain state of geomassive in the affected area of mine workings to be performed.

The developed software complex can be applied to solve the problems of mathematical modelling, linked with:

- Performance of computing experiments for evaluation of a geomechanical state in the vicinities of goaf;

- Forecast of geomechanical parameters of underground mining technology in the process of design documentation preparation.

The recommended requirements to hardware and software providing functioning of the program packages:

- processor – Intel Core i5;
- operative memory – 8 Gb;
- empty space on the hard disk – 10 Gb;
- operating system – Microsoft Windows XP or later versions;
- package of symbolic mathematics – Mathematica 9;
- package for launching software on other computers written in the package Mathematica - Wolfram CDF Player Pro.

References

[1] Tsvetkov A B, Pavlova L D and Fryanov V N 2015 *GIAB* **1** 365–370
[2] Pisrenko G S and Mozharovsky N S 1981 *The Equations and Boundary Value Problems of the Theory of Plasticity and Creep* (Kiev: Naukova dumka) p 496.
[3] Petukhov I M and Linkov A M 1983 *Mechanics of Pressure Bumps and Outbursts* (M.:Nedra) p 280
[4] Stavrogin A N and Protosenya A G 1979 *Plasticity of Rocks* (M.: Nedra) p 301
[5] Tsvetkov A B and Pavlova L D 2015 *Proc. of the Int. Conf. on High Technologies of development and Use of Mineral Resources* (Novokuznetsk) pp 121–125
[6] Tsvetkov A B and Pavlova L D 2016 *Proc. of the IV All-Russian Conf. on Modelling and the Information Technologies in Engineering and Social and Economic Systems* (Novokuznetsk: SibSIU) pp 175–178

IOP Conf. Series: Earth and Environmental Science **45** (2016) 012010 doi:10.1088/1755-1315/45/1/012010

Adjustment of automatic control systems of production facilities at coal processing plants using multivariant physico-mathematical models

V F Evtushenko[1], L P Myshlyaev[1], G V Makarov[1], K A Ivushkin[2] and E V Burkova[1]

[1] Siberian State Industrial University, 42 Kirova Street, Novokuznetsk, Russia, 654007
[2] Consolidated Company "Sibshakhtostroy", 9 Kuznetskoye Rd, Novokuznetsk, Russia, 654006

E-mail: evtushenko_vf@sibsiu.ru

Abstract. The structure of multi-variant physical and mathematical models of control system is offered as well as its application for adjustment of automatic control system (ACS) of production facilities on the example of coal processing plant.

1. Introduction

One of the most important problems in ACS development is their structural and parametric adjustment, for which, as a rule, special testing ranges, workbenches, as well as existing physical models of control objects, functioning as parts of control system of a physical model (CSPM).

In [1, 2] the features, conditions and procedure for application of such systems in the test tasks, adjustment and research of full-scale control systems are considered. The efficiency of the procedures offered here can be significantly increased by generalizing CSPM in case of control of multivariate physico-mathematical model of a full-scale control object – multivariate control system of a physical model (MvCSPM).

The controlled component in MvCSPM is a combination of (by analogy with full-scale mathematical [3]) physical model of the technological control object in the form of its information representation [4], and mathematical incremental model, functioning in the increments to data on changes of input and output effects of physical model.

Multivariance is created by connecting to a single functioning physical model any finite number of recalculation models using the above indicated combination, and, thus, form a finite set of simultaneously functioning physico-mathematical models of full-scale objects of any physical nature and, accordingly, control systems, but they and their mathematical incremental models should have a similar structure.

The main advantage of MvCSPM is the fact that its use will allow the multiple full-scale control systems (FCS) to be simultaneously explored, tested and configured, provided that each FCS and the corresponding to it control system of physico-mathematical model will be similar [5]. Another positive feature of the physico-mathematical modelling in comparison with the full-scale mathematical one is the possibility to change easily enough the technological operational modes of the physical model, which is not always feasible or is associated with high costs for full-scale objects.

In this paper we examine one of the possible procedures adapted to the solution of problems of simultaneous adjustment of the finite set of ACS output actions and states of technological units at a coal processing plant, which are an important component of automated technological complexes, largely determining efficiency of their operation.

The proposed option is based on multivariate physico-mathematical modelling using full-scale data, obtained with the help of existing monitoring systems, as well as modelling and full-scale data with the specified statistical properties formed by the multivariant generator in the form of a closed dynamic system (MV generator (CDS) [6]); similarity of control systems [7-9], control of their of similarity due to purposeful change of the dynamic properties of external influences transformation channel and the properties of these influences per se, including uncontrolled ones; search procedures.

2. Structure and description of a multivariant model control system

The integrated structure MvCSPM is shown in Figure 1, where the following notations are used:

■ – elements of the physical model of a control object;

➞ – material and energy impacts of a physical model;

→ – informational impacts;

– measuring units, including incomplete control and measurement errors;

– execution units, including the implementation errors of the control signals (commands);

□ – calculation units of data processing;

$Z=\{U;W:Y\}$ – generalized vector of impact, including, respectively, control, external and output effects of the physical model;

y_{red} – reduced to the output disturbance effects as an integral evaluation of uncontrolled disturbances, expressed on a scale of output effects;

superscripts "*R*", "*F*", "*C*", "*P*", "*PM*", and "*" mean, respectively, real, full-scale, compromise values of impact of the full-scale control object, impacts of its physical, physico-mathematical model and the given values;

subscripts "*red*" and "*l*" – reduced impacts and number of a physical model ACS;

the dotted line indicates the presence of other parallel contours connected to PM, the number of which equals to L.

The ACS structure shown in Figure 1 corresponds to the control system by deviation, the object of which is affected by uncontrolled disturbances.

The mathematical part of the system in general is represented by the relations

$$Y_l^{PM}=Y^P+\delta Y_{ul}^{PM}+\delta y_{redl}^{PM};\quad l=\overline{1,L}; \tag{1}$$

$$\delta Y_{il}^{PM}=\varphi_{il}^{PM}\left\{\delta U_l^{PM}\right\};\quad l=\overline{1,L}; \tag{2}$$

$$\delta U_l^{PM}=U_l^{PM}-U^F;\quad l=\overline{1,L}; \tag{3}$$

$$\delta U_l^{PM}=f_l^{PM}\left\{\varepsilon_l^{PM}\right\};\quad l=\overline{1,L}; \tag{4}$$

$$\varepsilon_l^{PM}=Y_l^{PM*}-Y_l^{PM};\quad l=\overline{1,L}; \tag{5}$$

$$y_{redl}^{PM}=y_{redl}^{PM*}-y_{redl}^{P};\quad l=\overline{1,L}; \tag{6}$$

$$\delta y_{redl}^{PM}=Y^P-Y_i^P-Y_w^P; \tag{7}$$

$$y_{redl}^{PM}=F\left\{y_{red}^{P};\alpha_l^*;\sigma_l^*\right\}, \tag{8}$$

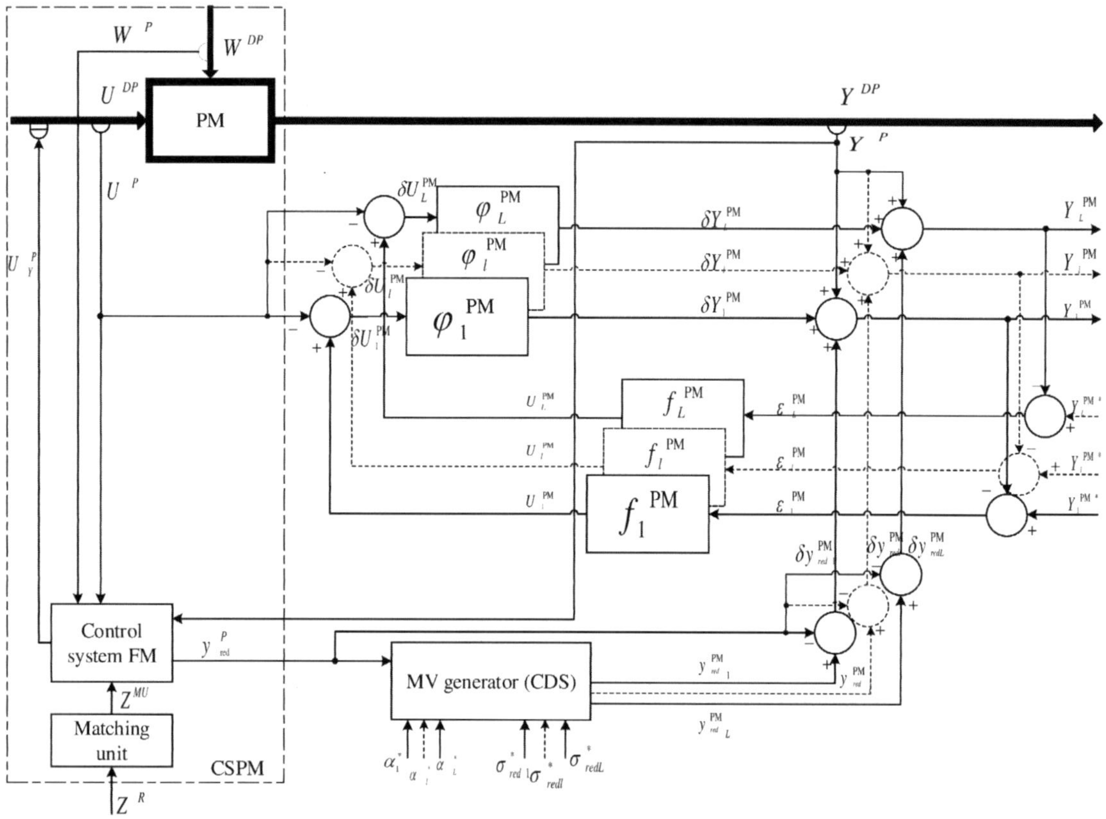

Figure 1. Multivariant model control system.

where $\varphi_l\{\}$ – operator of the mathematical incremental model l ($l=\overline{1,L}$, L – number of modeled ACS) physico-mathematical models of control object;

$f_l\{\}$ – control law of l ACS;

$F\{\}$ – operation algorithm of multivariant generator of time series data with given statistical properties;

Y_i^P – effects of regulatory impacts of the physical model.

Each (l in L) simultaneously working in MvCSPM physico-mathematical ACS should reflect dynamic properties and operating conditions of the corresponding l of the full-scale system, $l=\overline{1,L}$, which are generally different. To do this, l model and l full-scale ACS should be similar [1]. Their similarity is achieved in the mathematical part of the physico-mathematical model due to purposeful change of (ACS similarity control [1]) either the coefficients values of operator $\varphi_{il}^{PM}\{\}$, or properties of a time series reduced to the output of physico mathematical disturbance model δy_{redl}^{PM}, or both.

Formation of $\varphi_{il}^{PM}\{\}$ is carried out by changing the mathematical model of the channel transformation of regulatory impacts of a physical model $\varphi_i^P\{\}$, in particular through the use of special corrective operators $\varphi_{ic}\{\}$ [10]. These disturbances δy_{redl}^{PM} with the specific properties are formed using multivariant generator of time series data with desired properties, based on algorithm [6], adapted to the multivariant case.

3. ACS adjustment

The integrated scheme of full-scale ACS adjustment with the use of multivariate physical and mathematical models of the controlled system being the part of MvCSPM is shown in Figure 2. It was used for control system adjustment of the technological complex at the processing plant "Matyushinskaya". The description of technical realization of the complex is presented in [11].

Below, as an example we consider the application of this procedure for two ACS: the suspension density of a dense-media separator and dense-media hydrocyclone. Adjustment of the slurry density, fed to the separator and sump of hydrocyclones power supply, is performed in the ACS according to deflection by dilution of conditioned slurry with additional water. The structure of these full-scale and physico-mathematical ACS (Figure 1) is the same.

The *problem* is presented in the following form.

Given

1. The mathematical model of conversion channels of controlling impacts for dense-media separator and for dense-media cyclone has the form

$$\varphi(s) = \frac{k}{Ts+1} \cdot e^{-\tau s}, \tag{9}$$

where k – transfer coefficient, T – time constant of inertia, τ - time of pure delay.

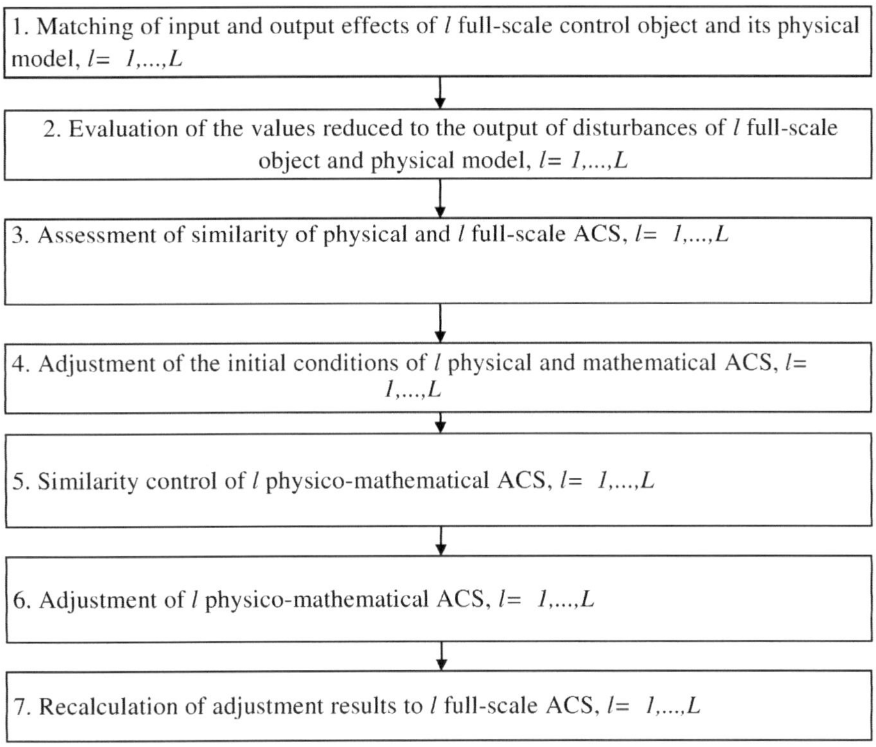

Figure 2. Scheme of full-scale ACS adjustment with application of MvCSPM.

2. Parameters of k, T, and τ models of control objects (dense-media separator and dense-media hydrocyclone), estimates of which are given within the ranges [11]

– for the dense-media separator

$$-2.3 \leq k \leq -1.8 \ \frac{kg \cdot m^3}{\%\ stroke};$$
$$7.0 \leq T \leq 12.0\ s;$$
$$5.0 \leq \tau \leq 7.0\ s; \tag{10}$$

– for the dense-media hydrocyclone

$$-2.0 \leq k \leq -1.4 \ \frac{kg \cdot m^3}{\%\ stroke};$$
$$7.0 \leq T \leq 12.0\ s;$$
$$5.0 \leq \tau \leq 7.0\ s. \tag{11}$$

The values of these coefficients were evaluated at the stage of design and further specified experimentally during implementation in the "local" and "remote" modes of operation.

3. The disturbances reduced to the output of full-scales objects, the estimates of which were calculated by the known pattern [12] on the basis of the initial data on the operation of full-scale ACS using average values of the coefficients k, T and τ from ranges (10) and (11).

The autocorrelation functions (ACF) of the reduced disturbances were approximated by the expression:

$$r_{red}(\theta) = \sigma_{red}^2 \cdot e^{-\alpha_{red} \cdot |\theta|}, \tag{12}$$

where σ_{red}^2 – dispersion of the reduced disturbance, α – ACF coefficient, θ – time of a shift between sections of a sequence of reduced disturbance. Average values of the reduced disturbances characteristics.

– for the dense-media separator

$$\sigma_{red} = 11.05(\frac{kg}{m^3}); \quad \alpha_{red} = 0.11; \tag{13}$$

– for the dense-media hydrocyclone

$$\sigma_{red} = 10.10(\frac{kg}{m^3}); \quad \alpha_{red} = 0.17. \tag{14}$$

4. Proportional-integral law of control with the transfer function

$$f_{pi}(s) = k_i + \frac{k_i}{s}, \tag{15}$$

where k_p and k_i – coefficients of its proportional and integral part.

5. Quality criteria of ACS operation in the form control mean-square errors

– for the dense-media separator: q_1^F and q_1^{PM};

– for the dense-media hydrocyclone q_2^F and q_2^{PM};

– for the physical model q^P,

where the superscripts "F", "PM", "P" means "full-scale", "physico-mathematical" and "physical" and subscript "1" and "2" – dense-media separator and dense-media hydrocyclone, respectively.

6. Multivariant physico-mathematical model of the control system – MvCSPM, the structure of which corresponds to the scheme in Figure 1. The basic controlled component of MvCSPM is a physical model of the production unit [13]. The conversion channel model of regulatory impacts of an object physical model corresponds to expression (9); its coefficient values k^P = 1.6 °C/min/%; T^P = 3.0 sec.; τ^P=1.0 sec. The reduced disturbances to the physical model output were calculated similarly to the full-scale object according to the scheme [12]. The normalized ACF, as well as the full-scale regulation objects, was approximated by expression (12) with the parameters values σ^P_{red} = 1.4 (°C); y^P_{av} = 1.1(°C); α^P_{red} = 0.3.

The control law of ACS model is described by expression (15), with the following values of its superscripts coefficients, corresponding to the minimum control mean-square error: k^P_p = 1.875; k^P_i = 0.625.

7. Multivariant generator of model and full-scale (data series) impacts with the given statistical properties, allowing the desired finite number of such data series to be simultaneously formed.

Required

To determine the values of superscripts coefficients k_p and k_i of proportional integral control law for full-scale ACS of slurry density of dense-media separator and dense-media hydrocyclone.

4. The scheme and solution results

The *solution* of the problem was carried out using MvCSPM, customized for ACS of production facilities, and the procedure of simultaneous adjustment of both physico-mathematical systems in accordance with the scheme shown in Figure 2 for $l=\overline{1.2}$.

1. Matching of the input and output effects of the full-scale control facility and physical model of disturbances was carried out in this example by their reduction to the dimensionless data by normalizing the expression

$$Z_F = \frac{Z - Z_{\min}}{Z_{max} - Z_{\min}}; \qquad Z = \{Y; y_{red}; U\}, \tag{16}$$

where $Z_{\max}$ и $Z_{\min}$ – the maximum and minimum values of the corresponding variables.

2. The calculation of the reduced to the output of considered full-scale objects and physical model of perturbations was made by a single scheme [12] for dimensionless data.

3. Similarity of full-scale ACS density slurry of a dense-media separator and hydrocyclone with the model ACS, the controlled component of which is a physical model, was evaluated by the correlation coefficient between the characteristics of and with respect to the proximity of targeted indicators reflecting the effectiveness of these systems [8]. As the latter the mean-square control errors were used.

As a result of estimation of these ACS similarities the following average sample estimations of characteristics were obtained:

$$q_1^F = 1.104; \quad q_2^R = 1.064; \quad q^P = 0.720; \; r_{pf1} = 072; \quad r_{pf2} = 0.65,$$

where r_{pf} and $r_{pf\ 2}$ – accordingly, the correlation coefficient between the q_1^F and q^P, q_2^F and q^P.

Comparison of the correlation coefficient values and performance characteristics of ACS operation with the required for their similarity [7-9] leads to the conclusion about the absence of similarity between the model and full-scale ACS. The values of the correlation coefficient between the mean square control errors of full-scale and a model ACS is significantly less than the required $r^{\bullet}_{FP}$ = 0.95±0.5. This conclusion is confirmed by the values $\delta q_1^F = \left| q_1^F - q^P \right|$= 0.387 и $\delta q_2^F = \left| q_2^F - q^P \right|$= 0.344, which is much higher than the permitted level of their deviation (10%).

Thus, the lack of similarity between the full-scale and model systems of ACS makes it necessary to deliberately change (control similarity) dynamic properties of the model ACS for establishing their similarity.

4. Adjustment of the initial conditions of physico-mathematical ACS was implemented via the following sequence of operations: 1) opening of the negative feedback of physical-mathematical ACS; 2) evaluation of deviations of regulatory impacts of full-scale and physical ACS; 3) recalculation of these deviations with the operator $\varphi^{PM}\{\cdot\}$ for calculation $\delta Y^{PM}\{\cdot\}$; 4) further in accordance with the scheme in Figure 1 the calculation of the output impact value $Y^{PM}\{\cdot\}$ of physico-mathematical system of regulation.

The implementation of the indicated above operations allows the value of the output impact at the output of ACS physico-mathematical model to be defined, corresponding to the value of its full-scale system.

5. Similarity control for both full-scale ACSs was carried out due to changes in the reduced disturbances properties with the help of MV generator (CDS), and due to changes in the operator parameters $\varphi^{PM}\{\cdot\}$. The results in the form of properties values of conversion channel models of regulating impacts and reduced disturbances of similar ACS are given in the table below. Figures (1) and (2) shown in the first row of the table in parentheses mean that all the characteristics and parameters related to the dense-media separators are marked in the text by figure 1, and dense-media hydrocyclones – by figure 2.

6. Adjustment of model ACSs was carried out in the mode of physico-mathematical simulation using search engine optimization techniques (deformable polyhedron [14]) by the minimum of mean-square control error. The following values of ACS adjustment for the dimensionless data on changes in input and output impacts of ACS physico-mathematical model were obtained.

– for the dense-media separator

$$k_{n1}^{PM} = 0.773; \; k_{i1}^{PM} = 0.011; \; q_1^{PM} = 0.871; \tag{17}$$

– for the dense-media hydrocyclone

$$k_{p2}^{PM} = 0.863; \; k_{i2}^{PM} = 0.01; \; q_2^{PM} = 0.883. \tag{18}$$

Table 1. Parameter values of ACS characteristics.

	Dense-media separator ($_1$)		Dense-media hydrocyclone ($_2$)	
	Full-scale data	Data of physico-mathematical modelling	Full-scale data	Data of physico-mathematical modelling
σ_{red}	1.00	1.00	1.00	1.00
α_{red}	0.09	0.54	0.10	0.50
k	-2.00	-1.60	-1.80	-1.60
T	12.00	2.00	11.00	2.20
τ	6.00	1.00	5.00	1.00

7. Recalculation of the parameter values of adjustment (17) and (18) for the condition of full-scale ACSs functioning was made in accordance with the expressions [1]

$$k_{P1} = k_{P2} \cdot \frac{k_1 \tau_1 T_2}{k_2 \tau_2 T_1} \cdot e^{\frac{1}{\beta}(\cdot \alpha_1 \cdot \tau_1 - \alpha_2 \cdot \tau_2)}; \tag{19}$$

$$k_{I1} = K_{I2} \cdot \frac{y_{av} k_2 \tau_2}{y_{av} k_1 \tau_1} \cdot e^{\frac{1}{\beta}(\cdot \alpha_1 \tau_1 - \alpha_2 \tau_2)}. \tag{20}$$

The values obtained for these parameters also for dimensionless data indicative of change in ACS effects are listed below under (21) and (23); in the second line under (22) and (24) the values of the same parameters in the natural scale of their change are given

– for the dense-media separator

$$k^{F}_{p1} = -0.618;\ k^{F}_{i1} = -0.001;\ q^{F}_1 = 0.870; \tag{21}$$

$$k^{F}_{p1} = -0.618 \frac{\%\ stroke \cdot m^3}{kg};\ k^{F}_{i1} = -0.001 \frac{\%\ stroke \cdot m^3}{kg \cdot s};\ q^{F}_1 = 9.613 \frac{kg}{m^3}; \tag{22}$$

– for the dense-media hydrocyclone

$$k^{F}_{p2} = -0.767;\ K^{F}_{i2} = -0.007;\ q^{F}_2 = 0.883. \tag{23}$$

$$k^{F}_{p2} = -0.767 \frac{\%\ stroke \cdot m^3}{kg};\ k^{F}_{i2} = -0.007 \frac{\%\ stroke \cdot m^3}{kg \cdot s};\ q^{F}_2 = 8.918 \frac{kg}{m^3}. \tag{24}$$

In the result of the use of ACS adjustment values (22) and (24) in the full-scale conditions and their further specification the conclusion about the high efficiency of adjustment procedure with the use of physico-mathematical modelling was made. Deviations of full-scale coefficient values from the model ones and the values of intermediate modulus error of control for these systems do not exceed 10%. In addition, the simultaneous execution of modelling experiments in the process of ACS adjustment allows the time and material-energy resources associated with the functioning of the physical model to be significantly reduced.

It is also important that the simulation modelling complex based on MvCSPM can be an effective tool in the learning process for the solution of problems connected with testing, adjustment and study of ACS.

5. Conclusions

The use of multivariant control systems of physico-mathematical models is appropriate for adjustment of the automatic control systems of production facilities, including for coal processing plants as well as in the training process of universities students and personnel of industrial enterprises and organizations. It is advisable to apply the proposed adjustment process of control systems that include more complex structures of production facilities, particularly facilities with recycling, distributed controls, etc.

6. Acknowledgements

The research is supported by the grant of the Russian Foundation for Basic Research, Project No. 15–07–01972.

References

[1] Myshlyaev L P, Ivushkin A A, Evtushenko V F et al 2015 *Proc. of Int. Conf. on High Technologies of Development and Use of Mineral Resources* (Novokuznetsk: SibSIU) pp 211–218
[2] Evtushenko V F, Myshlyaev L P, Ivushkin K A, Makarov G V, Burkova E V 2015 *Proc. of X All-Russian Conf. on Automation Systems in Education, Science and Production* (Novokuznetsk: SibSIU) pp 159–165
[3] Avdeev V P 1979 *Ferrous Metallurgy* **6** 131–134
[4] Myshlyaev L P and Evtushenko V F 2002 *Forecasting in Control Systems* (Novokuznetsk: SibSIU) p 358
[5] Myshlyaev L P, Evtushenko V F et al 2013 *Ferrous Metallurgy* **12** 40–42
[6] Myshlyaev L P, Evtushenko V F et al 2014 *Proc. of All-Russian* Conf. on *Information and Telecommunication Systems and Technologies* (Kemerovo: KuzSTU) pp 402–403
[7] Myshlyaev L P, Evtushenko V F et al 2012 *Bulletin of Siberian State Industrial University* (Novokuznetsk: SibSIU) **2** 41–43
[8] Myshlyaev L P, Evtushenko V F et al 2013 *Ferrous Metallurgy* **12** 58–60
[9] Evtushenko V F, Starovatskaya S N et al 2013 *Proc. of IX All-Russian Conf. on Automation Systems in Education, Science and Production* (Novokuznetsk: SibSIU) pp 32–38
[10] Myshlyaev L P, Evtushenko V F et al 2010 *Ferrous Metallurgy* **11** 65–67
[11] Grachev V V, Tsiryapkina A V et al 2013 *Proc. of IX All-Russian Conf. on Automation Systems in Education, Science and Production* (Novokuznetsk: SibSIU) 69–71
[12] Rotach V Ya 1973 Dynamics Calculations of Industrial Automatic Control Systems (M.: Energiya) p 439
[13] Emel'yanov S V, Korovin S K, Myshlyaev L P 2007 *Methods of Identification of Industrial Plants in Control Systems* (Kemerovo: Kuzbasvuzizdat) p 307
[14] Rykov A S 1999 *Methods of System Analysis: Optimization* (M.: Economics) p 255

Research into the impact of shafts misalignment of turbocompressor installation on power characteristics of a drive engine

M V Kipervasser and A V Gerasimuk

Siberian State Industrial University, 42 Kirova Street, Novokuznetsk, 654007, Russia

E-mail: kipervasser2012@yandex.ru

Abstract. In the present paper the emergency situation in operation of a turbocompressor installation is considered, which leads to shafts misalignment, connected by rigid couplings, of the drive synchronous electric motor and multiplicator is considered. Misalignment of shafts causes vibrations in the course of work of a turbine unit that, in its turn, can result in untimely wear of bearings, loosening of different fasteners and breakdown of the whole installation. The analysis of shafts misalignment of electric motor and planetary gearbox (multiplicator) regarding the emergence of mechanical power surge and derivation of corresponding equations is performed. The mathematical modelling of a turbine unit operation taking into account misalignment in MatLAB Simulink is carried out. The graphs of mechanical power surges on the shaft of the electric motor and current surges in the phase of its stator are received. The conclusion about the possibility of recording stator current surges by standard measurement methods and use of this fact for development of a current protection against misalignment of shafts is made.

1. Introduction

Turbocompressor installations are widely used at the mining and metallurgical enterprises. The compressor machines of centrifugal type, intended for pumping of large volumes of air or gas with a rather small pressure (0.15 – 1.0 MPa), are usually used. Turbocompressor installation usually is a high-speed and energy-intensive unit. Capacities of synchronous motors reach 12.5 MW, and rotation rate of turbocompressors wheels in some cases make 30000 rpm. Similar capacities and rotation rates condition high requirements to quality of installation and pre-commissioning activities.

One of the moments requiring especial attention is alignment procedure of turbocompressor and planetary gearbox shafts, and also - planetary gearbox and the electric motor. At high rotation rates the shafts misalignment leads to vibrations which negatively influence the whole installation. Vibrations result in the accelerated wear of bearings, loosening of bolted connections, failure of the connection couplings. Besides, during start-up of turbocompressor installations the emergence of the resonance mode of natural oscillation frequency and oscillations caused by shafts misalignment can lead to a severe accident.

Control of the installation vibrations is performed by means of vibration diagnostics. However, as practice shows, vibration control is not always applied. Sensors are a weak link of the monitoring system. Their failure paralyzes the system; it either becomes not operational, or loses its sensitivity. It is reasonable to use such monitoring system where the vibration sensors would be absent per se.

2. Calculation of additional load torque

For detection of shafts misalignment it is possible to use the indirect method of emergencies recognition based on current oscillations of the drive engine stator [1]. Misalignment of shafts creates some additional load torque on a shaft leading to the current oscillations of the drive engine stator which can be recorded [2].

In Figure 1 the emergence of shafts misalignment in the horizontal plane during assembly of the turbocompressor unit is shown. Shift in horizontal direction makes some value "s". In the initial point of time, before the beginning of the rotation, the shafts in the place of connection keep their the in-line state.

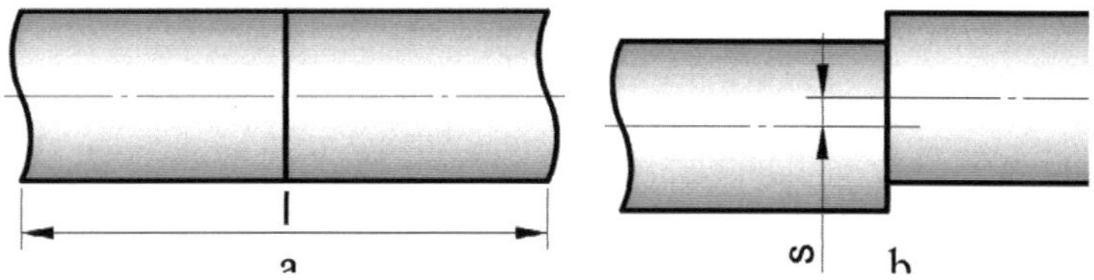

Figure 1. Position of the shafts The provision of shaft prior to rotation: a – side view, b – top view.

After energising the engine and the beginnings of rotation, due to misalignment of axial lines and under the influence of elastic forces, the rigidly connected shafts begin periodically bent at some value f (Figure 2). At the same time, the maximum value of the bend will be observed during rotation by 180 degrees.

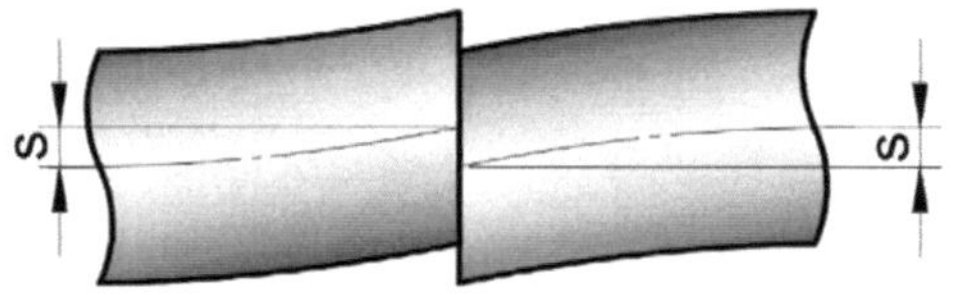

Figure 2. Position of the shafts during rotation by 180 degrees relative to the initial position. Top view.

The value of shafts deflection, with their lengths and diameters being equal, will be:

$$f = \frac{2s}{2}, \tag{1}$$

where f – the deflection value, m; s – misalignment value of axial lines of shafts, m.

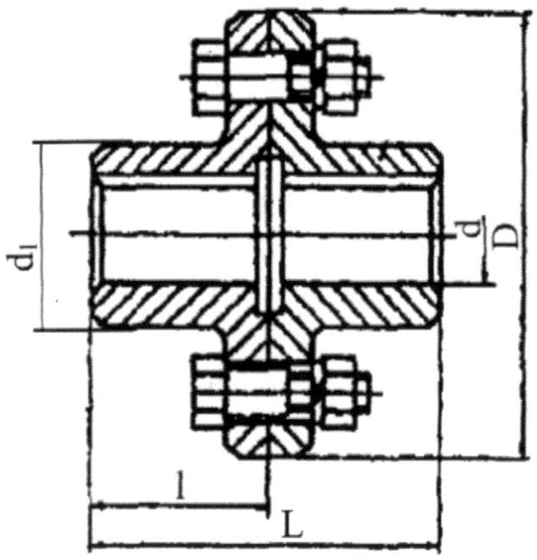

Figure 3. Drawing of a flange coupling.

It is necessary to consider that couplings are fitted over the shafts of the engine and planetary gearbox (it is accepted that the rigid flange couplings are used since when using other types of couplings, e.g., gear-type coupling, some misalignment of the shafts is compensated by the coupling design features) and the shaft as well as coupling bend (stretching of the connecting bolts is neglected). The drawing of the coupling is presented in Figure 3 [2].

The shaft deflection arises under the influence of bending force F defined as follows [3]:

$$F = \frac{48EIf}{L_f^3}, \tag{2}$$

where E – elasticity module equals $2{\cdot}10^6$kg/cm^2; I – equatorial moment of inertia of a solid shaft, m^4; f – shaft deflection, m; L_f – full length of couplings and a shaft (from the engine bearing to the reducer bearing), m.

$$L_f = L + 2(l_{48} - l_1 - 168) \tag{3}$$

where L = 250 – full length of the coupling, mm; l_{48} = 435, l_1 = 210 – the sizes according to the drawing of the engine, (Figure 4), mm.

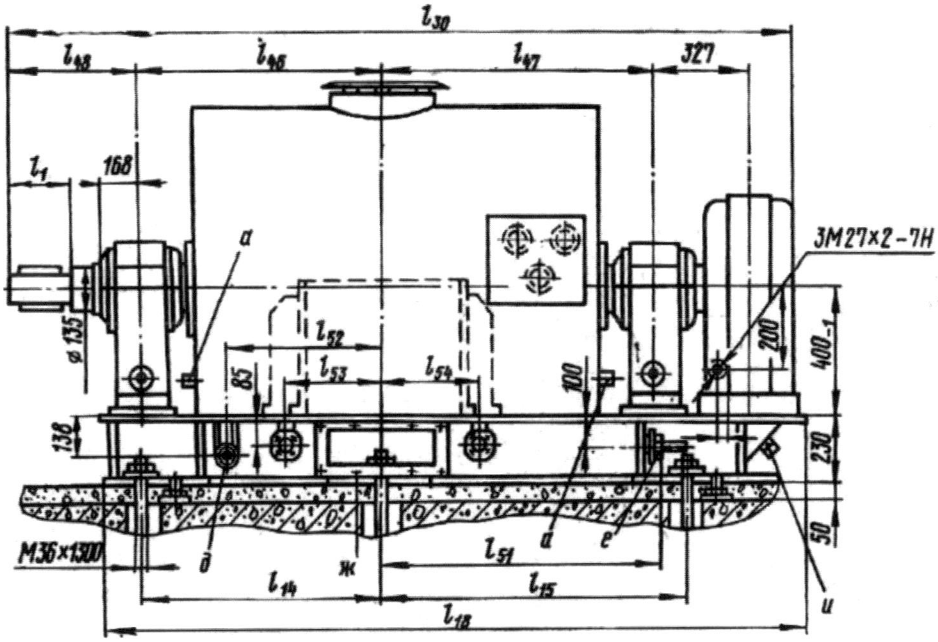

Figure 4. Drawing of the STD-1600 engine.

$$I = \frac{d_1^4}{64}, \tag{4}$$

where d_1 – diameter of the coupling, m (see Figure 3).

During shaft rotation the process of bending will have a periodic character and, obviously, will be described by function of a sine (provided that in initial position the shaft was in the non-bent state).

Then, taking into account formulas (2) – (4) it is possible to write down the bending force function on the shaft rotation angle:

$$F(\theta) = \frac{3Ef\pi d_1^4}{4L_f^{\ 3}}\sin(\theta), \tag{5}$$

where θ is a shaft rotation angle.

Dependence of a bending force on time is important for research. The variable dependent on time is the rotation angle:

$$\theta(t) = \omega t = 2\pi nt, \tag{6}$$

where n – shaft roation rate, rps; t – time, s.

Let us find the work, which the force F performs during the shaft rotation. To make one rotation the following work should be done:

$$A = \frac{d_1}{2}\int_o^\theta F[\theta(t)]d\theta(t). \tag{7}$$

The mechanical power, energizing the shaft and allowing it to perform one rotation, can be calculated by the formula:

$$P_M(t) = \frac{d_1 n}{2}\int_o^\theta F[\theta(t)]d\theta(t). \tag{8}$$

Taking into account formula (7) formula (8) will take the form:

$$P_M(t) = \frac{3Ef\pi d_1^5 n}{8L_f^{\ 3}}\int_o^\theta \sin[\theta(t)]d\theta(t). \tag{9}$$

After solution of the integral we will receive the following expression:

$$P_M(t) = \frac{3Ef\pi d_1^5 n}{8L_f^{\ 3}}\{1-\cos F[\theta(t)]\} = \frac{3Ef\pi d_1^5 n}{8L_f^{\ 3}}[1-\cos(2\pi nt)]. \tag{10}$$

Upon transition to dimension of rotation rate – rpm, formula (10) will take the form:

$$P_M(t) = \frac{Ef\pi d_1^5 N}{160L_f^{\ 3}}\left[1-\cos\left(\frac{\pi Nt}{30}\right)\right]. \tag{11}$$

Besides the power consumption for a periodic shaft bend, during the bend the losses of power due to bearing friction increase.

Dependence of friction force in the bearings on time has the form:

$$F_B(t) = z\mu F(t), \tag{12}$$

where z – number of bearings; μ – friction coefficient of bearings.

The power supplied for bearings friction is described by the expression:

$$P_B(t) = z\mu\frac{Ef\pi d_1^5 N}{160L_f^{\ 3}}\left[1-\cos\left(\frac{\pi Nt}{30}\right)\right]. \tag{13}$$

Then, total additional losses of power due to misalignment of two shafts will be:

$$P(t) = \frac{Ef\pi d_1^5 N}{160L_f^{\ 3}}\left[1-\cos\left(\frac{\pi Nt}{30}\right)\right](1+z\mu) \quad (14)$$

To obtain data on behavior of electric parameters of the drive electric motor in the situation of shafts misalignment the research of mathematical model was conducted. Mathematical modelling of operation of the turbocompressor installation being a part of compressor K-250-61-1 and electric motor STD-1600 with the planetary gearbox P1500/3,69 was carried out in the environment of MatLAB Simulink. Implementation of formula (14) by the standard blocks from Simulink library is presented in Figure 5.

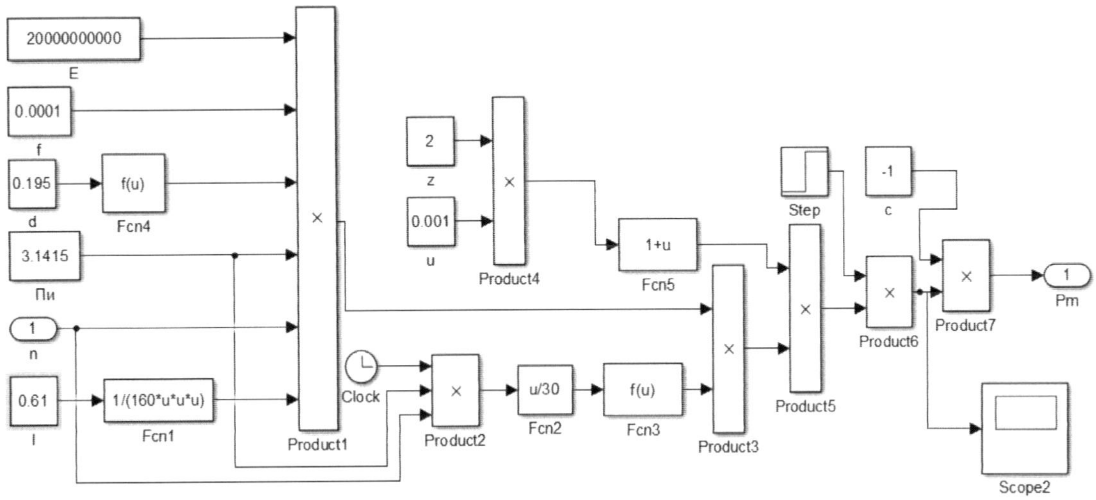

Figure 5. Functional scheme of imitation block of shafts misalignments.

The functional diagram of a model of the compressor unit, misalignment of shafts and the supply mains is given in Figure 6.

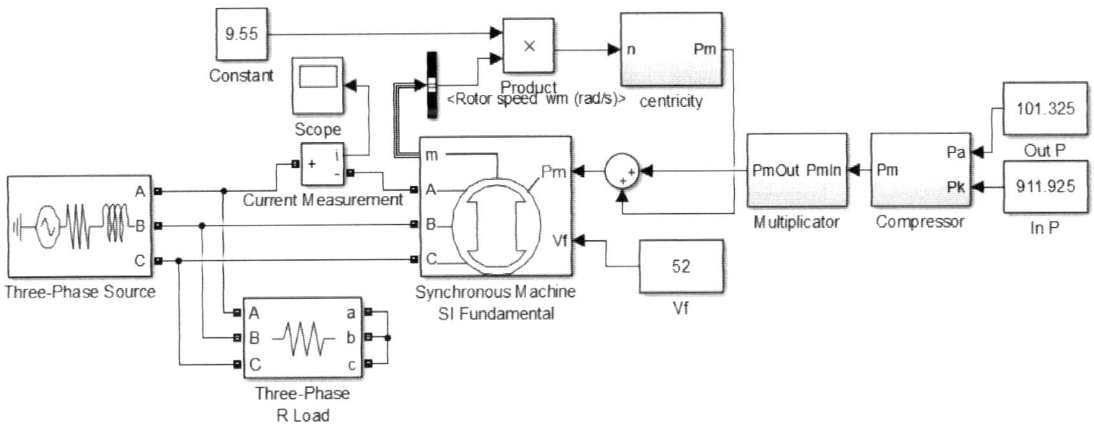

Figure 6. Functional structure of the mathematical model of the compressor unit and supply mains.

When modeling the shaft misalignment value of the engine and planetary gearbox was set 0.1 mm. The similar misalignments can be observed in real situation of turbocompressor installation functioning. The length of the free shaft end of the engine, according to the data of the producer, is 0.21 m, and diameter – 0.13 m. The diameter of the connecting couplings – 195 mm, and the full length of the coupling is 0.5 m. The parameters of the shaft of the planetary gearbox are accepted similar to the engine shaft. Modelling was carried out for the mode of nominal performance and air pressure [4]. The mode of misalignment arises on the third second of modelling.

3. Results of modelling

Figure 7 provides the current diagram of one of the electric motor phases during shafts misalignment, and Figure 8 shows the diagram of total additional losses of power due to misalignment.

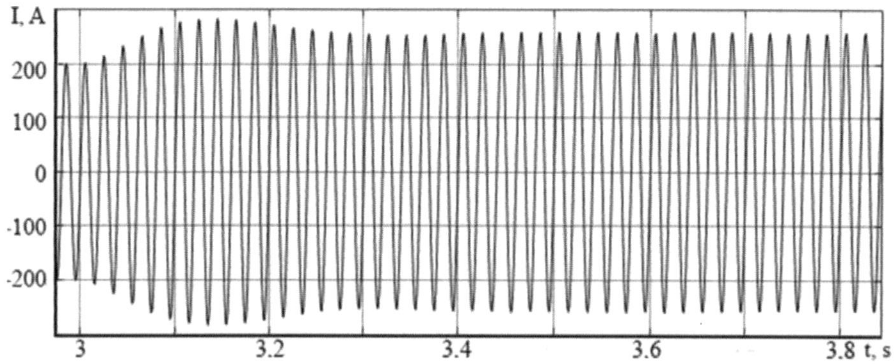

Figure 7. Current diagram in the phase of the drive engine during shafts misalignment.

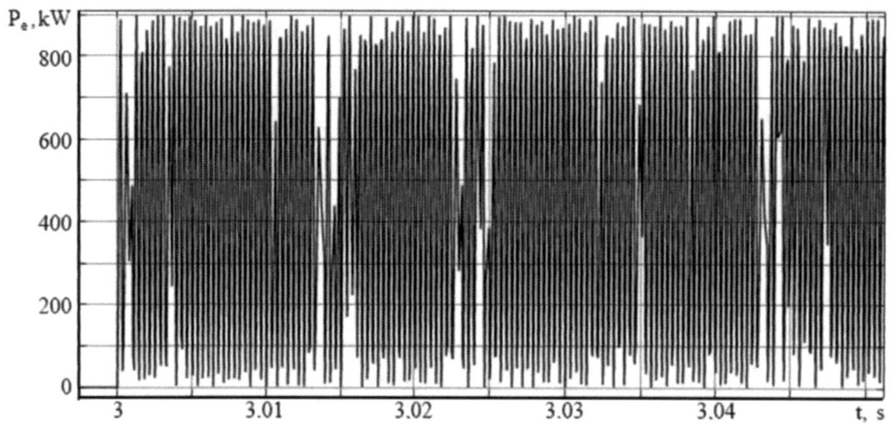

Figure 8. Diagram of mechanical power surge due to misalignments of shafts (the timescale is increased).

According to the diagrams the misalignment of shafts creates periodic (confirmed by the analysis of formula (13)) mechanical power surge with an amplitude up to 800 kW at the set parameters, that makes 50% of the engine nominal performance. Amplitude of the phase current surge of the electric motor equals to 290 A, that is 45% of the rated current and in the course of attenuation of transition process decreases to 255 A (28% of the rated current). Duration of transition process is 0.3 s.

4. Conclusions

On the basis of the obtained data it is possible to conclude that the value of current surge is sufficient enough to be registered by standard methods of measurement, and, therefore, it is possible to apply the method of indirect diagnosing of emergency situation [1] and to create the protection system against shafts misalignment as one of the components in the protection complex of turbine units.

References

[1] Pugachov E V, Kipervasser M V and Inzhelevskaya O V 2011 *Bulletin of KuzSTU* **4** 51–55.
[2] Pugachyov E V, Kipervasser M V and Gerasimuk A V 2016 *Bulletin of KuzSTU* **2** 84–90
[3] *Russian National Standard: Flange Couplings 20761-96* pp 2–11
[4] Postnikov I M 1960 *Design of Electrical Machines* (Kiev: GITL) p 274
[5] Sidorov M D 1962 Reference Book on Air and gGs-blowing (Moscow-Leningrad: Mashgiz) p 220

Adaptive optimization as a design and management methodology for coal-mining enterprise in uncertain and volatile market environment - the conceptual framework

V V Mikhalchenko[1] **and Yu T Rubanik**[2]

[1] T. F. Gorbachev Kuzbass State Technical University, 28 Vesennyaya Street, Kemerovo, Russia, 650000
[2] Center for Management Technologies, Po Box No. 13 Zelinograd, Moscow, Russia, 124617

E-mail: v.mikhalchenko@mail.ru

Abstract. The work is devoted to the problem of cost-effective adaptation of coal mines to the volatile and uncertain market conditions. Conceptually it can be achieved through alignment of the dynamic characteristics of the coal mining system and power spectrum of market demand for coal product. In practical terms, this ensures the viability and competitiveness of coal mines. Transformation of dynamic characteristics is to be done by changing the structure of production system as well as corporate, logistics and management processes. The proposed methods and algorithms of control are aimed at the development of the theoretical foundations of adaptive optimization as basic methodology for coal mine enterprise management in conditions of high variability and uncertainty of economic and natural environment. Implementation of the proposed methodology requires a revision of the basic principles of open coal mining enterprises design.

1. Introduction

Market economy conditions have significantly changed the characteristics of the business environment, criteria and strategy for ensuring viability of coal-mining enterprises of Russia. The ability of coal mine producers to meet demand for coal products and to provide high economic efficiency of invested capital are of fundamental importance in the market conditions.

For a majority of contemporary Russian coal mining enterprises the main barrier to achieve these objectives is a high degree of variability of business environment conditions. The variability of the business environment is evident in the seasonal and random changes in the demand and prices for coal product. Variability also affects the price and availability of material and human resources. Also, macroeconomic parameters changes add significantly to the variability of the market environment. As a consequence, uncertainty of future values of the parameters of the business environment is a fundamental and inevitable factor of market environment.

The second significant factor preventing the achievement of these objectives is a complete mismatch between majority of mines design parameters and their market environment. In a centralized planned economy, the positive effect of economies of scale was considered as a major source of economic efficiency. As a result a dominant strategy in coal mining production was a creation of big mining companies to exploit this effect to the full extent. Unfortunately, the bigger a scale of production system is, the higher inertia of that system is in a sense of time it takes to react to changing market conditions. Following this approach in the volatile market environment leads to significant economic losses. Historically creation of the finished coal products stock was considered as a basic mechanism for coal mining industry to protect it against market demand variations. This mechanism used to be well established and efficient in the planned economy environment. In the volatile market conditions it leads to excessive growth of working capital, economic risks increase, lower investment attractiveness and competitiveness of coal mines.

In recent years, the issue of coal-mining enterprises efficiency in the volatile market environment has been actively investigated by Russian scientists [1-5]. The aim is to achieve market competitiveness of coal mining companies and bring them to the level of efficiency corresponding to the global energy market. However, the issue is still under consideration and approach to develop coal mining system robust and adaptable toward market variation needs further research. The problem under consideration is of systemic nature. To resolve it – it is necessary to revise the principles and methods of production systems design, as well as organizational and managerial processes. All above mentioned proves a relevance importance of the research approach presented in this article.

The object of the study. The production system of the open pit coal mining, operating in a volatile and uncertain environment in the market economy.

The subject of research. Conceptual framework and design principles for coal mining enterprise operating in volatile market environment, including models and algorithms of adaptive control at the strategic and operational level of management in the short-term and long-term (investment) time frame.

Research methods: theory of adaptive systems and mathematical theory of optimization.

2. Results

The general concept of the necessary transformations of coal mines can be defined as the alignment of the dynamic properties of coal mining system with dynamic properties of the market environment. The implementation of this concept involves a change of the design principles of production systems, as well as changes in the organizational, economical and management systems of coal mining.

In the time frame of business planning cycle in order to decrease level of economic risk and to increase net cash flow at optimum stock level the production planning principles need to be changed. The leading principle in this respect should be the synchronization of the coal production rate with a variation of market demand. A necessary condition for transition to the synchronized mode of operation is the achievement of a low enough level of coal mine production cycle time. That means that production cycle time has to be less than time horizon for reliable demand prediction, and less than the time constant corresponding to maximum frequency in the power spectrum of market demand for coal product in the chosen market segment. These premises were defined in [7] as a condition of predictability and controllability for adaptive mode of production system operation.

In the frame of long-term corporate investment cycle the most important decisions are associated with target markets, placement, production capacity and dynamic parameters of newly build and modified mines. As a result, the whole corporate system of production and logistics units have to be optimized in order to increase the market share and profitability of the coal mining corporation, and reduce investment risks. Marketing and investment policies should guide development of production capacity and dynamic parameters of new mining enterprises according to the dynamic structure of demand at the target market.

The "lower frequency" part of market demand in power spectrum has to be served by mines with a high production capacity and lower dynamic properties – relatively long production cycle. As demand variability at this part of the demand power spectrum is relatively low, so, adaptation of this part of corporate production facilities should use coal stocks buffering mechanisms. Thus, corporate exploits the opportunity of economy of scale effect.

The "higher frequency" part of market demand in power spectrum has to be served by mines with a lower production capacity and high dynamic properties – short production cycles. Since this component of demand in power spectrum displays a high volatile behavior, excess production capacity is considered as the most relevant mechanism for market environment uncertainty adaptation. The economic rationale for reserve production capacity should be based on the comparison of the volume and cost of capital invested in the creation of a reserve power with the amount and cost of capital tied up in stocks of finished coal products and work-in-process.

To reduce the production rate and the level of investment risk within the life cycle of the enterprise, and increase responsiveness of the open pit mining production system to variations of the market demand, it is necessary to change its fundamental design principles. Systems of coal mining should be formed on the basis of adaptive phased design patterns, providing the ability to change production capacity and volume of economic resources used at any stage of the life cycle in accordance with changes in the market environment and, at the same time, reducing technology driven capital tied up in mining works for time periods far beyond the horizon of market size long term prediction.

3. Discussion

As it is known, the viability of the biological, social or economic systems is determined by their ability to adapt to changing environmental conditions. The adaptive system is a system able to modify its behavior algorithms, functional and material structure with the aim of maintaining or achieving optimal state when changes in the external environment occur [6].

Thus, the ability of mining companies to change their parameters or structure in accordance with the changes of the market environment at different stages of the life cycle, determines their viability, and in an economic sense, their efficiency and competitiveness.

Adaptation should be considered as a step-by-step process aimed at maximum possible increase of economic efficiency at each step based on achievement of cost leadership and the level of dynamism adequate to the variability characteristics of the mining and geological conditions and market environment. As it was shown in [7], a necessary condition for such type of adaptation process is the formation of capacity of the control object to modify its parameters or structure during a time period that do not exceed the time frame of significant changes in the natural and business environment within control cycle at all levels of organization system.

The lack of adaptation capabilities for majority of coal mining corporate systems in Russia manifests itself in the conflict between the indicators of economic efficiency and ability to sustain uncertainty and variability of external environment. This conflict is predetermined and embedded in the production system long before its manifestation – at the stage design of its life cycle due to major criteria used in its design. Until this time the prime concerns of designers at this stage are issues of final surface location of mining works, production capacity, parameters of the working area. As a consequence, the dynamic parameters of the production system are not considered and not taken into account when major decisions regarding choice of mining system and technological processes are taken.

Generally speaking, the entire conceptual mining framework developed in the conditions of centralized planning economy including design criteria, design methodology, social legislation limitation and policies define capabilities and economical and technical limitations of existing enterprises. In the radically different environment of market economy the design, technical, technological and management decisions taken within this obsolete framework produce evident and significant negative effects, create economical conflicts. These solutions have become a basic causal factor limiting the adaptability of coal mines to the characteristics of volatile and undefined external environment. One may consider these negatives effects as a clear and lucid signals – the prevailing system of design production and management in the mining industry has to be changed.

The fundamental premise laying at the foundation of traditional mining conceptual framework is a supposition of environment stability and predictability. As a result, a designer supposes that it is possible to rely upon the data regarding economic and natural environment parameters available at the moment of decision taking to make “optimal” decisions. These optimal decisions are considered to be valid at a current time, as well as over the period of the life cycle, the duration of which usually exceeds 50 years or more. Obviously, in a market economy for a specific period of time it is impossible to provide a reliable forecast of the economic environment parameters, including average market demand. Limited predictability and uncertainty parameters of the economic environment in the long term is the root cause of significant economic risks that affect the investment attractiveness of coal mines and determining the amount of deviation of the parameters from the optimum.

To reduce the economic risks, ensure the sustainable operation of enterprises throughout the life cycle it is necessary to overcome the current conceptual mining industry framework. That means that decision making processes at all stages of life cycle should meet the terms of adaptability. To this end while making design decisions one should consider external environment parameters as time dependent and known with acceptable accuracy only within a certain time horizon. These parameters should not be used for making decision beyond the horizon of their reliable prediction. Consequently, the system of coal mining should be designed so that it could change its parameters or structure in accordance with changes in the external environment. To ensure these conditions, the trajectory of development of the enterprise within the framework of the whole life cycle should be considered as a sequence of investment cycles. The duration of each investment cycle should be chosen in such a way that condition of predictability and controllability condition mentioned above to be fulfilled.

Obviously, traditional methods of making design decisions in mining industry do not meet these conditions. The source of the problem is a supposition of rigid spatial structural relationship of the open pit working space. This supposition is useful in a sense that it helps designer to define a fairly narrow limits for the trajectory of working space development during life cycle of open pit mine and set such parameters as a final depth of the pit to achieve maximum NPV – discounted net income over the entire life cycle of coal-mining enterprises. In turn, the decision made regarding the final borders location of a pit that will be reached over 50 years or more, determines and limits such crucial structural, technical and economic relationship for a distance of rock mass transportation, the dependence of the current stripping ratio, variable costs of the current and maximum depth of the pit.

It is worth mentioning that rigid spatial structure supposition which fixes location for the final borders, leads to a such trajectory of working space that maximizes operating costs and investment at the initial and middle stages of the life period. That means that return on these cost will be shifted into distant and uncertain future. This kind of spatial trajectory of working space predefines essential technology driven capital investment tied up during the major part of the whole span of mine life cycle in the hope to achieve economy at very distant time, for many decades of years to come. This approach to optimization could be justified in some degree in the stable and predictable economic environment of centralized planned economy and state ownership of the means of production. In the conditions of market economy, the values of many parameters of the economic environment, which are to be used to calculate the optimum value of the final depth of mining (volume of demand, inflation, interest rates, cost of capital, prices for the main types of resources and so on) for remote periods of time, are characterized by a high degree of uncertainty. As a result, there is a high degree of uncertainty with regard to the final depth of the pit and all the technical characteristics which are deducted from it, as well as structural proportions and economic parameters of the company.

Failure to comply with the terms of adaptability at the design stage determines the instability of the design decisions and the need for periodic essential design parameters correction during the reconstruction. However, the reconstruction implements and fine-tunes the parameters, but does not change the spatial structural relationships and enterprise development trajectory of the.

To assess the dynamic features of coal mining system in the timeframe of the whole life cycle it was proposed [7] to use the "index of adaptability". One may calculate that index as a ration of the time frame of changes in economic parameters of the external environment that significantly affect the nature of the design solutions to the time it takes to change some critical characteristics of the production system, such as production capacity, type of technology and its main parameters.

As analysis has shown [7], there are just a few environment parameters with essential influence on the character of the design decisions. They are: the level of market prices; volume of demand in the chosen market segment; the width of power spectrum of the market demand; the cost structure of the major resources used by the coal mining enterprise.

In a stable market environment the period of significant changes to these parameters may arise in a time span of 7-10 years; for volatile environment this period of relative stability may shrank to 2-3 years and even less than that.

Technology, for which the value of adaptability is more than one, should be classified as flexible, i.e. capable of adaptive behavior. Indeed, the technology for which the specified index is less than one, should be characterized as rigid, with a low level of adaptability.

Taking into account current and projected fundamental shifts in the patterns of global energy consumption there is no wonder that the horizon for reliable forecast of technical and economic parameters that determines the choice of optimal trajectory for enterprise development shrinks all the time. As a result, a rigid nature of most open pit mine enterprises, which use deepen-longitudinal development systems, became apparent [Figure 1a, b].

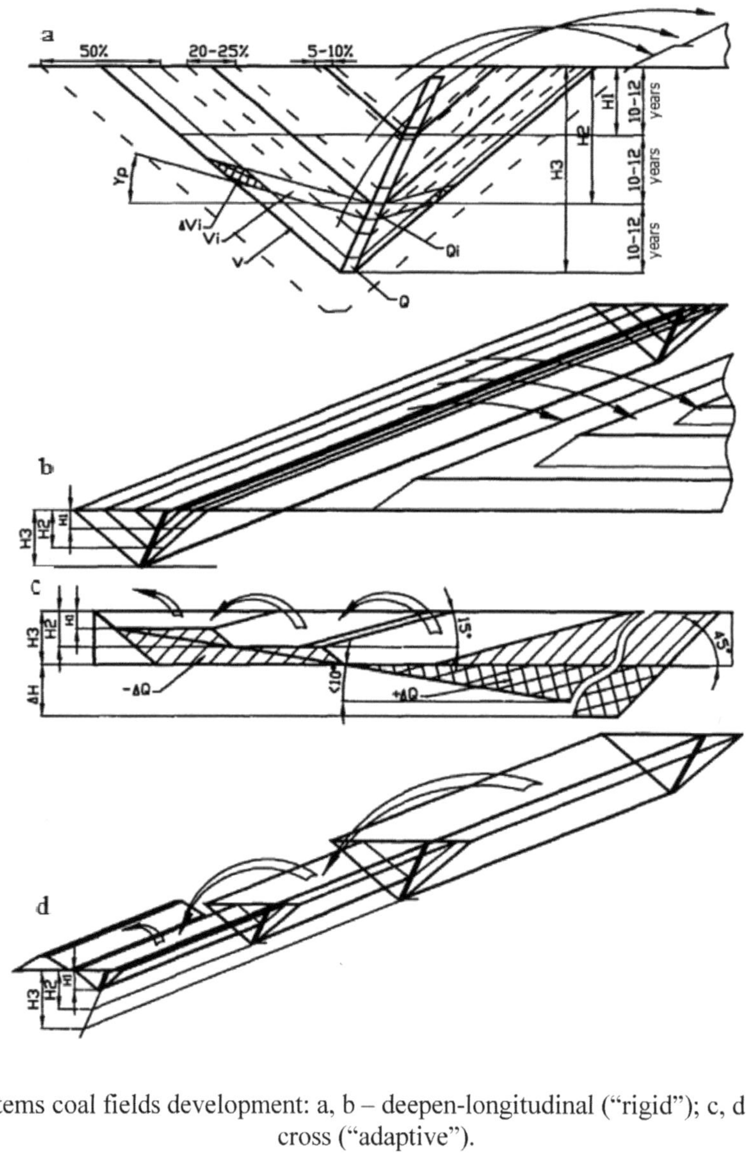

Figure 1. Systems coal fields development: a, b – deepen-longitudinal ("rigid"); c, d – deepen-solid cross ("adaptive").

As the coal-mining enterprises cannot change the dynamic characteristics and long term trends at the energy markets and reduce the uncertainties in the external business environment, then the only effective way to satisfy conditions for adaptability is the development and implementation of design solutions that break the chain of cause-and-effect relationships, which form the "rigid" coal mining conceptual framework.

A radical solution of this problem is a transition to phased, flexible technologies of mining operations [Figure 1c, d]. These technologies may be considered as ones which to a great extent fulfill and idea of adaptive behavior and the mentioned above criteria of predictability and controllability. The basic idea behind these technological solutions is to increase the number of degrees of freedom in the choice of possible directions of development of the working space of the open pit mine, and reduce the transport distances of the rock mass. For many coal mining companies an adequate choice might be, for example, mining technological system, referred to as "group deepen-solid cross", for which the spatial-temporal characteristics of each stage can vary in accordance with changes in the characteristics of the business environment [8, 9].

The determination of the direction and parameters of mining operations in this case is carried out at each successive stage of enterprise development. Duration of stages, during which the parameters of the mining system remain unchanged, are determined not by the inertia of the enterprise but by the level of instability and uncertainty of the natural and market environment over time. As a consequence, the enterprise avoids technologically defined capital tie-up in the mining works, and investment into questionable economical benefits to be gain in the future periods beyond the horizons of reliable prediction.

An ability to change the parameters of the production system in accordance with the significant changes in the business environment delivers evident economic and competitive benefits: characteristics of the enterprise on each segment of trajectory is sufficiently close to the optimum so that the maximum value of cash flow at the current settings of economic and natural environment is achieved.

Essential advantage of the "group deepen-solid cross" technology is an essential reduction for distance of rock mass transportation. This results in a significant improvement of the dynamic characteristics of the production system, as well as radical reduction of the cost of production during most of the trajectory segments. The duration of the production cycle in the same geotechnical conditions, when using this and similar technological systems, can be reduced by several times in comparison with the currently widely used deepen-longitudinal development system.

Such dynamic capabilities gives a possibility to radically change the mode of coal mining enterprise operation and switch to so-called "synchronized" or "just in time" mode. This mode of operation means that stock buffers all along production-logistic chain of mining industry are reduced to minimum with a corresponding reduction of capital tied up in a finished products and work in a process. Analysis of demand in Kuzbass regional energy market has shown that the synchronized mode of operation might be used to satisfy more than 95% of potential demand.

Studies have shown that the proposed use of "flexible" technologies also allows the rate of resources consumption by the producers to be reduced significantly (2-3 times), which would increase the production efficiency of open-pit coal mining and, as a result, would enhance its resource potential.

4. Conclusions

Thus, the transition to adaptive optimization as a basic methodology to design and manage open pit mine enterprises gives them a possibility to survive and prosper in volatile and uncertain economic environment of postindustrial era. Flexible models for open pit mines working space development allows mining companies to achieve a significantly higher productivity and production efficiency and, consequently, get competitive edge in the fierce competition at the world energy market.

It should be noted that to great degree all the benefits of transition toward new design and control methodology are still in the area of opportunities, of possible but not real future. To make these benefits real, significantly higher level of dependability for all processes in production, logistic has to be achieved. New, more sophisticated and robust methods for coordination of production and supporting units and processes at all organizational levels and control cycles need to be developed and applied. Higher labor competence and discipline, employee involvement and cooperation are required. Therefore, a transition to a new and higher level of management should be accomplished.

References

[1] Malyshev Yu N, Zaidenvarg V E et al 1996 *Restructuring of Coal Industry: Theory. Experience. Programs. Forecast* (Moscow) p 536
[2] Malyshev Yu N and Zykov V M *Russian Coal Industry in XXI Century* http://www.rosteplo.ru/Tech_stat/stat_shablon.php?id=272
[3] Zaidenvarg V E, Podolyak V I et al 2003 *Analysis of the Impact of Globalization on Mineral Resources* (M.: Institute of Economic Strategies) p 180
[4] Groeneveld B and Topal E 2011 *J. of Mining Science* **2** 212–226
[5] Mihalchenko V V 2008 *Proc. of the V China – Russia Symp. on Underground Engineering and Building of City and Mine* (China, Qingdao, 26 – 28 Sept.) pp 41–44
[6] Ashby U 2006 *An Introduction to Cybernetics* (M.: URSS) p 432
[7] Mikhalchenko V V 2002 *Organizational and Economic Basis for the Formation of Adaptive Systems in Coal Mining abstract of* Dr. Ec.Sc. thesis (Chelyabinsk) p 35
[8] Mikhalchenko V V et al 1992 *Method of Development of the Inclined and Steep Deposits with Great Thickness Along the Strike* A.C. No. 1745941, appl. No. 4822675/03 (033899)
[9] Mikhalchenko V V and Atrashkevich A A 1999 *Method of Open-pit and Underground Development of the Inclined and Steep Deposits with Great Thickness Along the Strike* patent of Russia No. 2143075

Management of local economic and ecological system of coal processing company

T V Kiseleva[1], **V G Mikhailov**[2] **and V A Karasev**[2]

[1] Siberian State Industrial University, 42 Kirova Street, Novokuznetsk, 654007, Russia
[2] T. F. Gorbachev Kuzbass State Technical University, 28 Vesennyaya Street, Kemerovo, 650000, Russia

E-mail: mvg.eohp@kuzstu.ru

Abstract. The management issues of local ecological and economic system of coal processing company – coal processing plant – are considered in the article. The objectives of the research are the identification and the analysis of local ecological and economic system (coal processing company) performance and the proposals for improving the mechanism to support the management decision aimed at improving its environmental safety. The data on the structure of run-of-mine coal processing products are shown. The analysis of main ecological and economic indicators of coal processing enterprises, characterizing the state of its environmental safety, is done. The main result of the study is the development of proposals to improve the efficiency of local enterprise ecological and economic system management, including technical, technological and business measures. The results of the study can be recommended to industrial enterprises to improve their ecological and economic efficiency.

1. Introduction

The development of coal industry determines the high level of anthropogenic impact on the environment, causing the need to resolve the problem of resource and nature conservation. A special role belongs to the coal processing enterprises, which, by controlling the quality of processed products, create conditions to improve the environmental safety of all the natural, technical, and ecological and economic systems for the extraction, processing and use of coal [1]. The coal processing plants also make all kinds of negative impacts on the environment [2, 3]. The improvement of efficiency of nature conservation activities depends on the local ecological and economic system "Concentration plant – environment" management process. The additional complication is caused by the uniqueness of the particular ecological and economic system and labor content of use of existing management algorithms developed for the other similar objects [4, 5]. Therefore, the study of the properties of ecological and economic systems and management methods requires the independent research in every particular case. This situation is complicated by such property of most ecological and economic systems, as the irreversibility of processes occurring in them. On this basis, we can formulate the main objectives of ecological and economic improvement of coal preparation process:

- environment protection from all kinds of man-made impacts resulting from the coal preparation plant activities and the use of the processed coal products;
- resources conservation and rational use of natural resources by reducing losses of coal, waste recovery, and increasing of the output and value of coal concentrates.

IOP Conf. Series: Earth and Environmental Science **45** (2016) 012013 doi:10.1088/1755-1315/45/1/012013

2. Materials and methods

The study object is the management system of ecological and economic system of coal processing plant. The study subject is LLC "Coal-processing Plant "Listvyazhnaya" ecological and economic system management efficiency based on selected indicators. This company has various negative effects on the environment. To study this process the statistical processing of environmental data on the basis of the relevant reporting forms was carried out. The elements of system analysis and the results obtained by specialists in the field of ecological and economic system management are also used in the paper.

3. Results and discussion

In the conditions of region with advanced coal processing industry the efficient management of local ecological and economic system "Industry – environment", the simplified management scheme of which is shown in Figure 1 [8, 9], is of vital importance [6, 7].

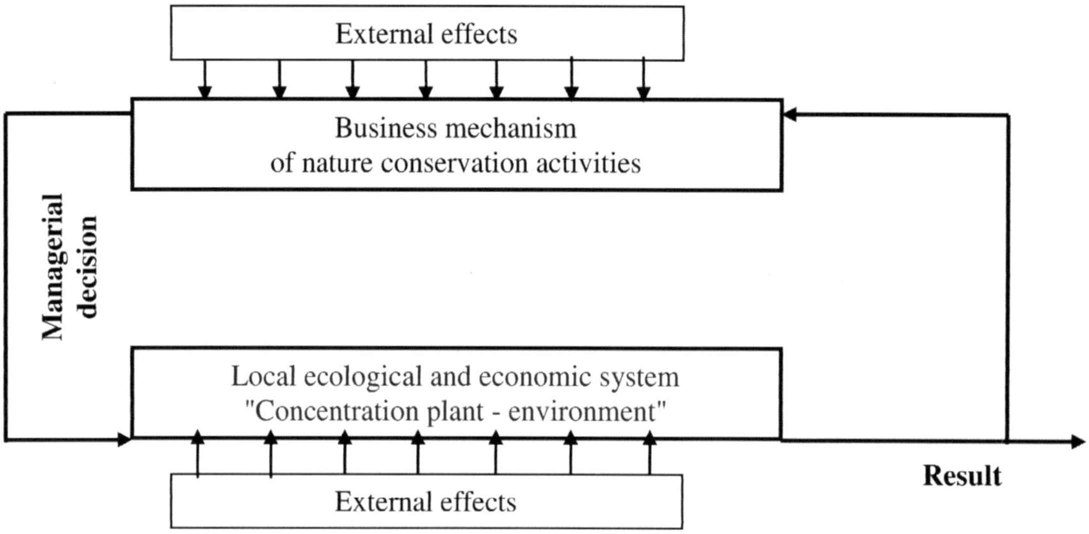

Figure 1. Simplified scheme of the local ecological and economic system management mechanism.

Figure 1 shows that a key element of the scheme of the local ecological and economic system management mechanism is the business mechanism of nature conservation activities [10-15]. The efficiency of the system functioning is evaluated by groups of related indicators presented in Figure 2.

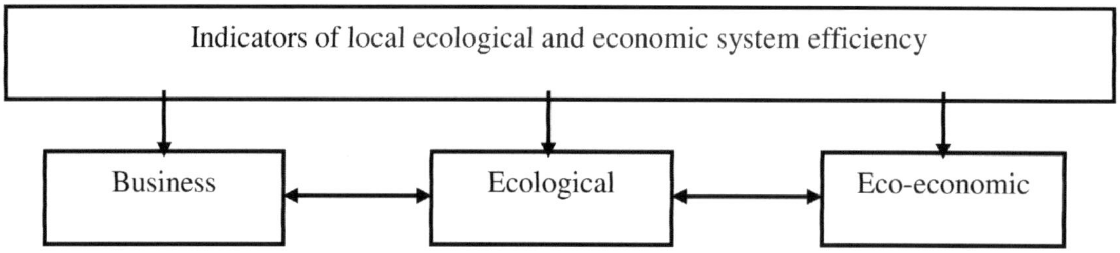

Figure 2. Correlation of the indicators functioning of local ecological and economic system efficiency.

The economic indicators of ecological and economic systems of coal processing enterprises include the run-of-mine coal processing products (output, concentrate, screenings, waste), profit, cost, technical condition of fixed assets, the level of capacity utilization, and others. The ecological indicators [16] include the parameters of negative impact on the environment (air emissions from stationary and mobile sources, sewage discharges, and industrial and consumer waste) and other characteristics. As seen in Figure 2, the determining indicators are eco-economic, including the cost valuation of the negative impact on the environment, as well as its reduction (economic assessment of the negative impact on the environment - economic damage; fee for the negative impact on the environment; fines for violation of environmental legislation, current expenditure on fixed assets intended for nature protection purposes, etc.).

To assess the efficiency of the local ecological and economic system we propose to analyze the structure and the dynamics of run-of-mine coal processing products, as well as the indicators fee for the negative impact on the environment, which is especially important in connection with the prospective increase of this payment [4, 5].

To test the research proposals the data of processing plant "Listvyazhnaya", which was commissioned in 2007 and became part of the Holding Company "SDS-Ugol" in 2010, were used. LLC "Coal-processing Plant "Listvyazhnaya" is a modern and powerful Russian coal processing plant involved in the steam coal processing which produces the high-quality D-grade coal concentrate (low ash and sulfur content, high caloric value) on an industrial scale.

This company pays great attention to environmental safety using the modern automated process and production management system. In addition, the company uses the modern coal processing technology and has a closed production cycle, using mine water. The coal processing waste are used for remediation of man-made landscape of the former mine "Inskaya". The utility water is purified with the modern treatment plants of Heat and Water Complex which applies the efficient physical and chemical purification method.

The assessment of the efficiency the local ecological and economic system of the processing plant can be carried out with the help of economic indicators (processing, output, concentrate, screenings and waste), shown in Table 1 and Figure 3.

Table 1. Structure of run-of-mine coal processing products of LLC "Coal-processing Plant "Listvyazhnaya".

Indicator	Measurement unit	2008	2009	2010	2011	2012	2013	2014
Processing	thousand tonnes	2800	3100	3500	4450	2900	4300	5400
	%	100.00	100.00	100.00	100.00	100.00	100.00	100.00
Output	thousand tonnes	2400	2500	2850	3750	2050	3100	4350
	%	85.71	80.65	81.43	84.27	70.69	72.09	80.56
Concentrate	thousand tonnes	1400	1300	1450	1900	1100	1800	2550
	%	50.00	41.94	41.43	42.70	37.93	41.86	47.22
Screenings	thousand tonnes	1000	1200	1400	1850	950	1300	1600
	%	35.71	38.71	40.00	41.57	32.76	30.23	29.63
Waste	thousand tonnes	400	600	650	700	850	1200	1050
	%	14.29	19.35	18.57	15.73	29.31	27.91	19.44

Table 1 allows us to conclude about the structure of run-of-mine coal processing products of coal processing plant. If we consider an indicator such as the "Output", it has a maximum value in the processing volume in 2008 (85.71%) and minimum in 2012 (70.69%). The important characteristic is the specific amount of the concentrate, which also has a maximum value in 2008 (50.00%) and minimum in 2012 (37.93%). The significant fluctuations in the proportion are observed when considering the "Screenings" indicator: from the minimum in 2014 (29.63%) to the maximum in 2011 (41.57%). The evaluation of the processed coal quality and the efficiency of management of the local ecological and economic system "Concentration plant – environment" can be made using the specific value of the indicator "Waste", which has a wide range of values from 14.29% in 2008 to 29.31% in 2012.

Improving the adequacy of evaluation of economic parameters of ecological and economic system can be achieved by studying the dynamics of indicators of run-of-mine coal processing products, presented in Figure 3.

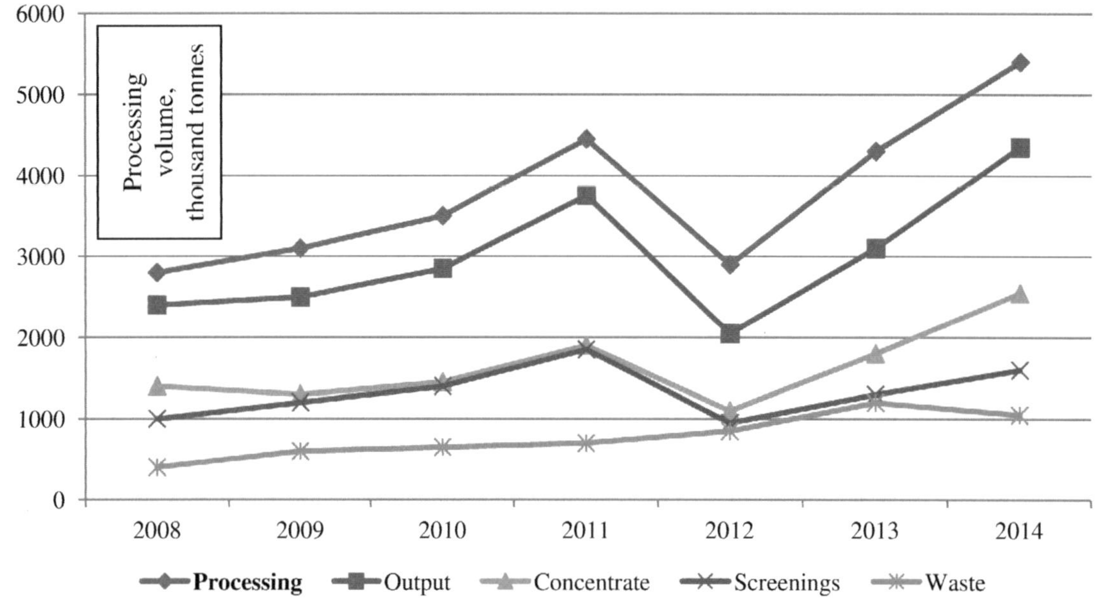

Figure 3. Dynamics of indicators of run-of-mine coal processing products of LLC "Coal-processing Plant "Listvyazhnaya".

Figure 3 shows that the indicators "Processing", "Output" and "Concentrate" follow the similar trend – the increase up to 2011, the decline in 2012 and rise to a new level in 2013 and 2014. Another positive trend is shown by the indicator of "Screenings" – after a peak in 2011 (1850 thousand tonnes) and minimum in 2012 (950 thousand tonnes) the growth rate slowed down to 1600 thousand tons in 2014. The dynamic of the indicator "Waste" is smoother – the sustained growth until 2013 (1200 thousand tonnes) and a significant reduction in 2014 (1050 thousand tonnes).

Despite the efficient environmental performance, the company LLC "Coal-processing Plant "Listvyazhnaya" makes various types of negative impacts on the environment, carrying out emissions of inorganic dust, iron oxide, manganese compounds, gaseous fluorine compounds, and produces the I, III-V hazard class waste. Therefore, the tightening of environmental legislation [6], including the prospect of increasing the raising factors in the calculation of charges for the negative impact from 5 to 100 requires the assessment of complex ecological and economic indicators of the local ecological and economic system "Concentration plant – environment", presented in Table 2.

Table 2. Indicators of the negative impact on the environment of LLC "Coal-processing Plant "Listvyazhnaya".

Environment element	2008	2009	2010	2011	2012	2013	2014
The fee for a negative impact on the environment within the established limits, including, kRUR:							
water	-	-	-	-	-	-	-
air	0.40	1.27	1.40	6.68	1.77	7.54	7.21
waste	-	-	-	1.81	0.01	0.22	0.53
Total	0.40	1.27	1.40	8.49	1.78	7.76	7.74
The fee for a negative impact on the environment over the established norms, including, kRUR:							
water	-	-	-	-	-	-	-
air	-	-	-	329.94	180.07	93.95	93.90
waste	37.60	2.63	2.65	15.76	-	66.15	48.60
Total	37.60	2.63	2.65	345.70	180.07	160.10	142.50
Total fee for a negative impact on the environment, kRUR							
	38.00	3.90	4.05	354.19	181.85	167.86	150.24
The specific fee for a negative impact on the environment per 1 thousand tonnes of run-of-mine coal processing, rubles / thousand tonnes							
	13.57	1.26	1.16	79.59	62.71	39.04	27.82
The specific fee for a negative impact on the environment per 1 thousand tonnes of concentrate produced, rubles / thousand tonnes							
	27.14	3.00	2.79	186.42	165.32	93.26	58.92

Table 2 shows that the maximum charges for the negative impact on the environment are observed in 2011 - 354.19 thousand rubles. The similar trend is observed for the specific fee for negative environmental impact per 1 thousand tons of run-of-mine coal processing (79.59 rubles / thousand tonnes) and specific fee for negative environmental impact per 1 thousand tonnes of concentrate produced (186.42 rubles / thousand tonnes). If we consider the total fee for a negative impact on individual elements of the environment, the maximum value for the air is observed in 2011 (331.71 thousand rubles) and for the waste – in 2013 (66.37 thousand rubles). If we consider the maximum spread of rates of specific fee for a negative impact on the environment per 1 thousand tons of run-of-mine coal processing products and per 1 thousand tonnes of concentrate produced, for the first specific indicator ranges from rubles / thousand tonnes in 2010 and from 1.26 rubles / thousand tonnes in 2009 to 79.59 rubles / thousand tonnes in 2011. The second measure also has the minimum value in 2010 and 2009 (respectively, 2.79 rubles / thousand tonnes and 3.00 rubles / thousand tonnes) and a maximum in 2011 (186.42 rubles / thousand tonnes).

To improve the efficiency of management of the local ecological and economic system of the enterprise the technological and business solutions [17-25] aimed at the reduction of waste proportion and the increase of concentrate proportion in the run-of-mine coal processing products are required. In addition, there is a steady trend of high fees for the excess negative environmental pollution that requires a review of permits and the inventory of pollution sources. It should be noted that the current conditions for enterprises, including those actively operating in the foreign market, are connected to the need to use the environmental management systems, certification of products and processes according to the ISO-14001 series of standards. Another independent solution aimed to the increase of efficiency of the local ecological and economic system management can be the use of the best available technologies to reduce the environmental burden on the environment that is caused by the modern environmental requirements.

4. Conclusions

The concluded study of efficiency of the local ecological and economic system of the enterprise LLC "Coal-processing Plant "Listvyazhnaya" to improve the mechanism to support the management decisions, led to the following conclusions:

- the simplified scheme of the of the local ecological and economic system management mechanism is formed;
- the relationship of indicators of efficiency of the local ecological and economic system is shown;
- the structure of the efficiency indicators for the local ecological and economic system is studied;
- the analysis of the structure of run-of-mine coal processing products is conducted;
- the indicators of negative impact on the environment are considered;
- the technological and business solutions to improve the efficiency of the local ecological and economic system management are developed.

5. Acknowledgements
The authors express their gratitude to the management of the enterprise LLC "Coal-processing Plant "Listvyazhnaya" for the information provided.

References
[1] Kleyn M S and Vakhonina T E 2014 *Bulletin of KuzSTU* **3** (103) 125–128
[2] Galanina T V 2009 *Mining Informational and Analytical Bulletin* **7** (12) 174–177
[3] Kovalev V A, Potapov V P and Schastlivtsev E L 2013 *Monitoring of the Natural Environment of Kuzbass Coal-mining Areas* (Novosibirsk: SB RAS) p 312
[4] Burkov V N, Novikov D A and Shchepkin A V 2008 *Ecological and Economic System Management Mechanisms* (M.: Physical and Mathematical Literature) p 244
[5] Lipenkov A D 2005 *Modelling of Ecological and Economic Systems* (Chelyabinsk: CSU) p 130
[6] Pantyukova M E 2011 *Herald of KSTU* **17** (1) 200–202
[7] Skryl O K 2015 *Bulletin of KuzSTU* **5** 222–227
[8] Kravets O Ya and Solomakhin O N 2013 *Economy and Management of Management Systems* **3.1** (9) 190–198
[9] Mikhailov V G, Kiseleva T V and Mikhailov G S 2015 *Proc. of the Russian Conf. on Information and Telecommunication Systems and Technologies* (Russia, Kemerovo) URL: http: www.sibscience.ru/page/ITSIT/1-informacionnye-sistemy-v-nauke/1082.pdf.
[10] Busygin V P, Puzanov Yu V, Batyukhnova O G, Zakharova P V and Semenova I V 2012 *Ecological Systems and Devices* **4** 3–38
[11] Mikhailov V G and Bugrova S M 2015 *In the World of Scientific Discoveries* **6.1** (66) 548–564
[12] Mikhailov V G, Koryakov A G and Mikhailov G S 2015 *Journal of Mining Science* **51** (5)
[13] Bolsunovskaya Y A and Bolsunovskaya L M 2015 *IOP Conf. Series: Earth and Environmental Science* **24** 012003
[14] Nevidimova O G, Yankovich E P and Yankovich K S 2015 F *IOP Conf. Series: Earth and Environmental Science* **24** 012020
[15] Shinkevich A I, Lubnina A A, Koryakov A G, Mikhailov V G and Vodolazhskaya E L 2016 *International Review of Management and Marketing* **6** (2) 328–332
[16] Ponomarenko D, Tanayants I and Grunicheva S 2013 *TekhNadzor* **2** 30–31
[17] Vakhonina T E, Kleyn M S and Gorbunkova I A 2009 *Bulletin of KuzSTU* **1** 15–17
[18] Vaisberg L A, Azbel Yu I, Arsentiev V A and Kononov O V 2013 *Ore Dressing* **4** (346) 47–49
[19] Arkhipov V A, Paleev D Yu, Patrakov Yu F and Usanina A S 2014 *J. of Mining Science* **50** (3) 587–594
[20] Kondratev S A, Rostovtsev V I, Bochkarev G R, Pushkareva G I and Kovalenko K A 2014 *J. of Mining Science* **50** (5) 959–973
[21] Petukhov V N, Kubak D A and Subbotin V V 2015 *Coke and Chemistry* **58** (3) 101–108
[22] Lavrinenko AA and Golberg GYu 2015 *Coke and Chemistry* **58** (9) 362–366
[23] Lavrinenko AA. and Golberg GYu 2015 *Coke and Chemistry* **58** (10) 404–409
[24] Lavrinenko AA and Golberg GYu 2015 *Coke and Chemistry* **58** (11) 451–454

[25] Antipenko L A 2015 *Coal* **12** 68–71

IOP Conf. Series: Earth and Environmental Science **45** (2016) 012014 doi:10.1088/1755-1315/45/1/012014

Construction of a conceptual model of transport system for a coal mining region

Yu D Pristupa[1], V N Fryanov[2] and L D Pavlova[2]

[1] LLC "Consolidated PTU of Kuzbass", 13 Kirova Ave, Leninsk-Kuznetskiy, Russia, 652518

[2] Siberian State Industrial University, 42 Kirova Street, Novokuznetsk, 654007, Russia

E-mail: LNKPTU@suek.ru

Abstract. The methodological approaches to creation of a conceptual model of complex transport system for coal-mining region are substantiated. In the structure of the system base model the ensemble of local interconnected subsystems is distinguished. The local structure of the traffic management system of cargo transportation company is developed, the factors and indicators affecting the efficiency of cargo management are highlighted.

1. Introduction

The need for dynamic development of cargo transportation companies (CTC) in Kemerovo region is confirmed be the fact that the processes taking place in the complex transport system, are described by a large number of parameters, interaction of which is not possible to describe by the corresponding equations and relationships, that can be solved analytically.

The transport system in Kemerovo region corresponds to all the attributes of a complex system, that is proved by the following features [1]:

- unique character in terms of ensurance of coal flows – more than 200 million tons per year, while similar transport systems in other regions of Russia are characterized by a smaller scale of transportations and diversity of traffic flows of goods;
- weak structuring of theoretical and factual knowledge about the future of the transport system in connection with the reduction of coal production and capacity of coal market under the influence of growth of competitive energy sources, the requirements to the reduction of environmental pollution, expansion of pipeline transportation of oil and gas;
- composite character of the system, as it includes local interconnected subsystems in the form of cargo transport companies, distributed in the space of geological-economic regions and hierarchically subordinated to different firms, holding companies and industrial associations.
- heterogeneity of subsystems according to coal sources (mines, open-pit mines, processing plants), types of carbon storages, schemes of transportation flows on the private tracks;
- contingency and uncertainty of factors in the transport system, including the dynamics of coal market, seasonality of coal consumption, accidents at coal mines which are coal suppliers, climate conditions, failure of technical facilities, human factors, etc.;
- multicriteriality of the effectiveness evaluations of organizational and technological processes that result from the individual evaluation criteria of each subsystem performance; a

multiplicity of indicators characterizing the performance of the whole transport system; the presence of non-formalized qualitative indicators taken into account by the decision-maker;

- great dimension of the transport system, which requires the construction of an ensemble of local models for each sub-system or a group of sub-systems. In the study of the mutual influence of the set of subsystems the assessment of various aspects of the functioning of the entire system is carried out.

Despite individual characteristics of the transport system the common indicators can defined, that identify it as part of the theory of complex organizational systems [2]: the diversity in accordance with the physical and mathematical features; high adaptability of subsystems models and the entire system to real specific conditions; the acceptable risk of realization of control actions; multicriteriality of performance evaluation of each system element; large dimension and a multiple connections of subsystem elements, etc.

According to [3-4] when considering the transport system as an object of research a set of interrelated economic, social, natural and technogenic factors in the structure of the coal-mining region should be considered. The main actors of the transport system of the mining region are mining and raw materials processing companies: coal open-pit mines, ore and sand pits, mines, enrichment and sintering plants (Figure 1). The transport private tracks (technological roads in the structure of the enterprise) are intended for acceptance of mining enterprises products, storage, sorting, loading and shipment of products to consumers. On the public tracks of JSC "Russian Railways" (JSC "RZD") the formation of routes and goods shipment to customers takes place. The common structural subject of the transport system is the market of ready products. The transport system operates in a particular environment, which is formed with regard to the influence if nature and reactions of society in the form of operating personnel.

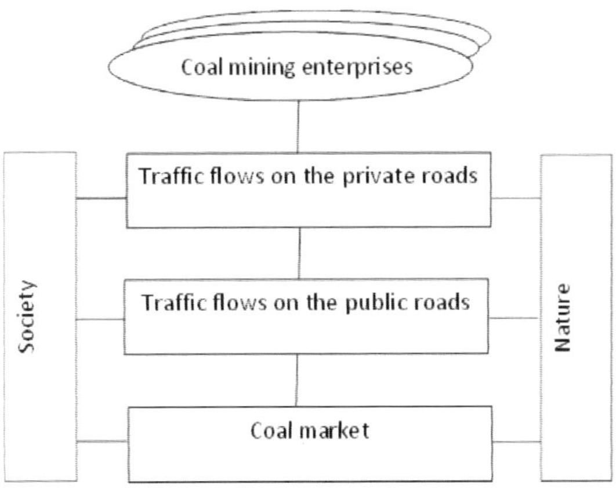

Figure 1. Subjects of the transport system in the mining region.

The mechanism of transport system functioning in the coal-mining region comprises a set of procedures and rules to be observed by the subjects when making management decisions. A special case is a control mechanism that includes only a set of managerial decision-making procedures. Mechanisms of functioning and management – strategy of behavior and decision-making by the parties.

The governing body (center) selects a decision-making procedure. The function of the governing body consists in prediction of the transport system development through modelling, selection and implementation of rational variants of the functioning of the control object with the subsequent monitoring and evaluation of management efficiency.

2. Construction of a conceptual model for a complex transport system

The most complete information about the complex transport system can be obtained by the experiment. However, the duration of the experiment with the developing system can be compared with the existence period of the researched object. In the conditions of current production the possibility for a pure experiment is also limited by the presence of internal and external influences on the object. In these cases it is necessary to construct and examine a model of a complex system.

The essence of modeling as a research method consists in the construction and analysis of models adequate to the transport system in the mining region. Using the model it is possible to analyze the response of the control system, to select and implement in practice specific control action, which leads to the fulfillment of the goal set by the control governing body (center).

Taking into account the possibility for forming a plurality of transport system models in the coal-mining region, it is advisable to take some basic model as a base, to which alternative models can be created.

The following variant, given in Figure 2, is offered as a base model.

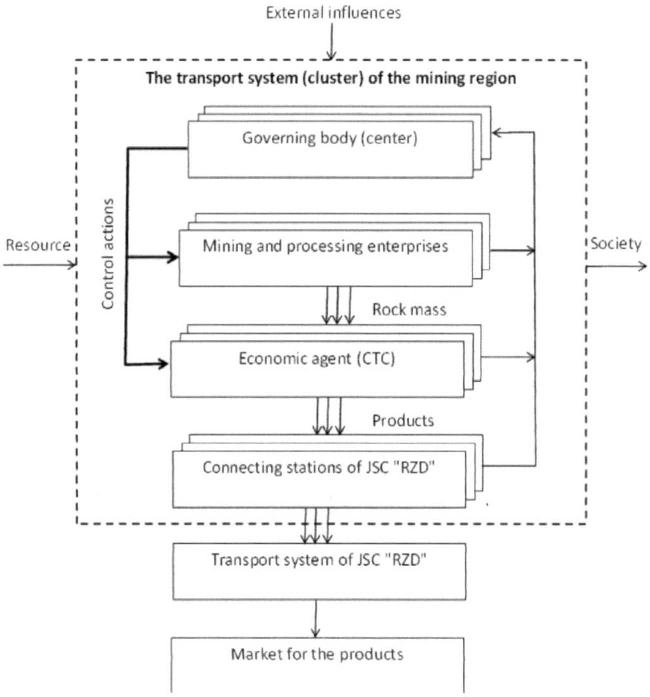

Figure 2. Structure a base model of the transport system in the mining region.

Holdings and large firms are included into the base model as governing bodies (centers), the production base of which is represented by mines, quarries, open-pit mines, processing plants engaged in the extraction and processing of mineral raw materials. The products of these companies are formed in the warehouses in the form of mined coal, ore, concentrate.

The presence of multiple elements of a complex transport system, such as mining and processing enterprises, roads, means of transport, coal storages, crushing and screening facilities, loading points, rail transport, tracks of non-public use, parks, stations, etc., predetermine the conditions of uncertainty. These conditions of the complex systems functioning do not allow the efficiency of the whole system to be evaluated by a separate isolated element or subsystem. Only an analysis of all subsystems with regard to their interaction can lead to a comprehensive evaluation of the entire system. However, in real conditions it is impossible to carry out the experiment due to the following reasons:

- high production costs during due to change of operational mode of the researched object in the process of the experiment in case the results of innovations are not satisfactory;
- duration of the experiment, which may correspond to the duration of the operation period of the facility.

In this regard, the analysis of the complex system can only be done through the design and analysis of a real object model. The solution of problems of a complex research object, as a rule, is carried out by construction and analysis of several interrelated local models that reflect different features of the complex system behavior as well various levels and stages of its examination.

To construct a working version of the base model of the transport system in the coal mining region it is necessary to establish the following local models in the structure of the basic model (Figure 2):

- transportation of the mined rock from mining and processing enterprises to the CTC storages;
- unloading, formation of storages, separation of rock mass, fire-safety measures;
- loading of the rock mass into the railway wagons;
- transportation of the rock mass in the wagons to the connecting station of JSC "RZD";
- shipping goods by freight rail to the consumer.

All local models should be divided into two ensembles: the external models and the internal models. The internal model are related with the simulation of activities of a particular cargo transportation company on the private tracks with the implementation of management mechanisms of enterprise state dynamics in the conditions of uncertainty [5]. The external models include descriptions of irregular freight traffic loaded with the mined rock coming from coal mining and processing enterprises and models of sales in the market environment with regard to the workload of JSC "RZD" tracks.

It should be noted that although the CTC subsystems interact within a single system, they have their own goals that may by achieved applying different criteria. In this regard, the main task of modelling and transport system management is the search for managerial decisions, which would harmonize local objectives and criteria of subsystems or eliminate contradictions between them.

The use of the base and local models for the modelling of alternative trajectories of object development will allow the formalized strategies and scenarios to be justified, implementation of which is carried out with participation of experienced specialists.

The composite nature of a complex transport system M involves representation of its model as a tuple of three groups of components:

$$M = \langle A, S, T \rangle, \tag{1}$$

where A – set of system elements; S – set of possible links between elements; T – a plurality of time points.

In the process of modelling it is possible to change the composition and structure of the transport system according to the realized t_i in a point of time.

Heterogeneity of subsystems requires the use of different mathematical methods: for example, for the description of coal acceptance from producers and formation of coal storages the no-queue models can be used, for the description of traffic flows on the private tracks the apparatus of differential equations with restrictions is often used.

When considering an ensemble of interacting subsystems, the cataloged library of models should be used, taking into account the scope and functions of each model.

In accordance with the structure of the base model of the transport system in the mining region the local structure of traffic control system for one of the main economic agents – cargo transportation company (CTC) was developed (Figure 3).

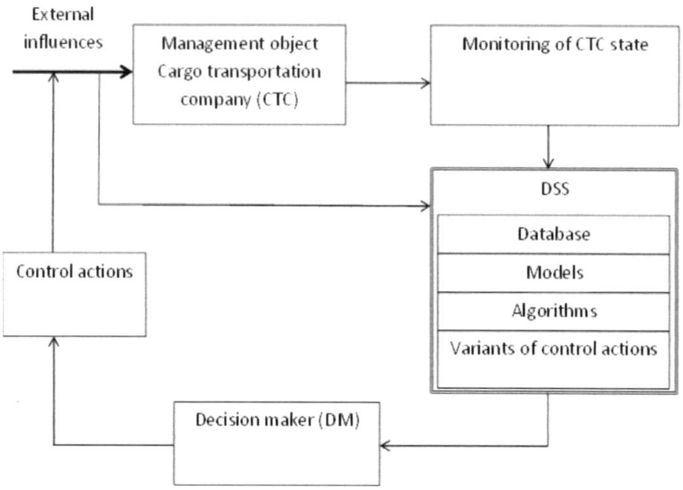

Figure 3. Control system of CTC cargo flows.

According to Figure 3 CTC as a control object functions in the structure of coal holding company or firm and carries out its activities taking into account external constraints in the form of scope of shipment to coal railway storages by coal producers, capacity of private and public tracks, coal market capacity, etc. Technological and organizational activities of CTC are effectively controlled, and the received the results of monitoring are formed in the form of a database. The efficiency of the results of the real technological and organizational processes is evaluated by the results of their modelling, and the variants of actions on the object are chosen in order to ensure the target indicators.

The final decision on the choice of the optimal variant of the action on the object is regulated by the decision-maker in the decision support system (DSS). While studying the problem situation the professional concepts and knowledge of the transport system researcher are used, which performs in the LLC "Consolidated Production Transport Directorate of Kuzbass" functions of a person making decisions.

To create a coal flow control model, which are formed in the coal warehouses for loading into wagons, transportation of the rock mass and the transfer of coal products to the consumer, it is suggested these processes to share the object of management, taking into account the functions performed, the following three subsystems (Figure 4):

I - processes of formation of a railroad coal depots dump unloading, processing and storage of the rock mass, loading, weighing and measuring cars; loading and unloading operations are carried out in-house and contracting means. Direct management of loading and unloading operations at the railroad warehouse workers responsible exercise - master loading complexes served by enterprises. In the production of handling operations involved in the processes of technological and organizational change in leadership personnel and daily regulations;

II – technological transportation of the rock mass by the private tracks, including shunting operations, selection of wagons with regard to shippers and consignees; wagons spotting to the places of loading and unloading; accumulation of wagons to be delivered to the connecting station;

III – acceptance operations by means of accumulation of railroad trains and their submission to the JSC "RZD" distribution network. The shipper routes of the specified weight and length are formed in accordance with the current plan of the West-Siberian Railway – branch of JSC "RZD" for formation of freight trains (in the direction of Avtovo, Vysotsk, Nakhodka, Tomusinskaya, Zherebtsovo, Novokuznetsk – North, Murmansk, Inskaya). The cargo and commercial work on the private railway track of JSC "SUEK-Kuzbass" PTU "Eastern Kuzbass" and the connecting station Terentevskaya is performed twenty-four-hour.

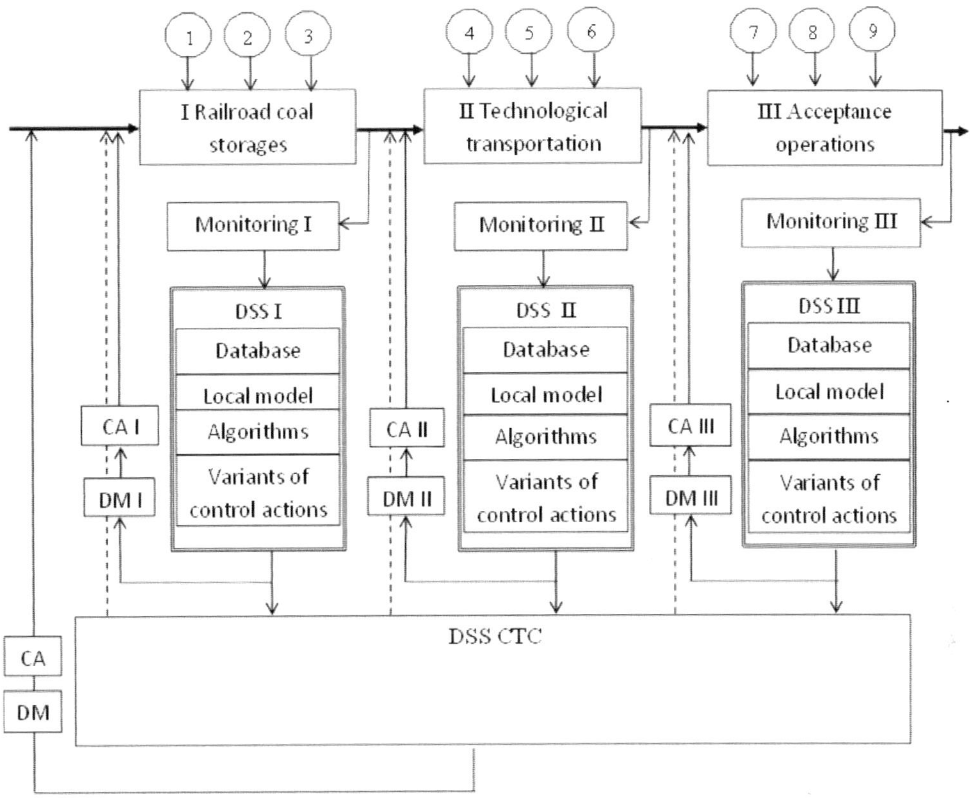

Figure 4. Structure of local elements and system models of freight traffic management at the CTC.

The following factors and indicators influencing the management efficiency of cargo flows at the CTC:

- scope and delivery rate of rock mass by coal mining and processing enterprises to the railroad storages;
- capacity of coal storage;
- processing capacity of loading points;
- state of the private railways;
- state and performance of locomotive transport and fleet;
- traffic management, communications, signalling arrangement;
- capacity of railways, stations to form trains;
- processing capacity of stations in JSC "RZD" network;
- capacity of the coal market.

Besides these main factors weather and seasonal conditions, human error, accidents and incidents, being the qualitative characteristics of uncertainty of the environment, influence on the efficiency of cargo flows control at CTC [6].

The identification of local subsystems in the structure of cargo flows control will improve the efficiency of decisions made by the relevant services and actors while controlling and adjusting the control actions (the doted lines in Figure 4).

3. Conclusions

Thus, a conceptual model, that contains the list and the characteristics of the interacting elements, processes and factors affecting the management efficiency of cargo flows at a CTC, was developed.

An obligatory component of interaction of elements and the ensemble of local models is a unified information richness at the individual level of persons making decisions at all levels of the hierarchy.

References

[1] Avramchuk E F et al 1988 *System Simulation Technology* (M.: Mechanical Engineering) p 520
[2] Burkov V N et al 2008 *Mechanisms of Management of Ecological and Economic Systems* (M.: Physical and Mathematical Literature) p 244
[3] Novikov D A 2007 *Theory of Control of Organizational Systems* (M: Fizmatlit) p 584
[4] Pristupa Yu D et al 2014 *Modelling of Process of Traffic Control at Coal Enterprise under Conditions of Uncertainty: Monograph* (M.: Scientific Review) p 208
[5] Pristupa Yu D et al 2015 *Proc. of Int. Conf. on High Tech Mineral Resources Development and Utilization* (Novokuznetsk: SibSIU) pp 195–200
[6] Shishkina S V et al 2015 *Proc. of X All-Russian Conf. on Automation Systems in Education, Science and Production* (Novokuznetsk: SibSIU) pp 166–173

On the problem of vehicles assignment in the conditions of open-cast mining

K S Davkaev[1], **M V Lyakhovets**[2]

[1] Synergo Soft Systems LLC, 91a Stroiteley Ave, Novokuznetsk, 654005, Russia
[2] Siberian State Industrial University, 42 Kirova Street, Novokuznetsk, 654007, Russia

E-mail: kdavkaev@sinergo.ru

Abstract. The problem of optimal truck-excavators assignment is considered. A new objective function is offered, which takes into account the performance of a given amount of work and the capacity of each truck and excavator. Different solutions are offered and an example of a problem solution is provided.

1. Introduction

The basic technological processes that accompany the open-cast mineral deposits development are: mining, capping (including organization and performance of blasting works), ancillary work (including re-excavation, trimming, stacking, etc.), reclamation. Almost all of these works are carried out with the use of excavators. Often mined minerals and rocks are removed by means of auto truck transportation. Of course, there are non-transport technologies and railway transportation. However, in the conditions of open-casts mining in Kuzbass Coal Basin the motor transport of coal and rock is used most often.

Given the need to reduce production costs, the solution to the problem of optimal trucks assignment for transportation of extracted minerals and overburden rocks will improve the economic efficiency of the mining enterprise. Moreover, it should be understood that the planning and management of technological production processes are now reaching a new level of responsibility – decisions taken on the basis of professionalism and engineering intuition, are ineffective because they do not take into account a number of opposing factors. To solve such problems it is necessary to use formal methods which can be well implemented with the help of computers.

2. Optimization criterion

Problems of this type belong to the class of allocation tasks and are solved by the use of linear programming methods. In [1-3] several approaches to solution of this class of problems in mining industry are considered. They are based either on the strategy of minimization of the cost for implementation of the specific scope of work, or on the strategy of maximization of the performed amount of work using the available equipment. However, in practice, the situation is different – it is needed to execute the planned volume of mining, overburden or other work during the the planning period (for example, one shift or one month).

Deviation from the specified scope of work can be taken as an optimization criterion:

$$\Delta W = \sum_{j=1}^{M} \left| W_j^* - \sum_{i=1}^{N} k_{ij} \cdot W_{ij} \right| \to \min \tag{1}$$

where: W_j^* – the planned work load for one month for the j-th excavator, j – type of an excavator, [m^3/month];

W_{ij} – the work load of the i-th truck during the work with the j-th excavator, i – type of the truck, $i \in N$,[m^3/month], and is defined by the formula:

$$W_{ij} = k_i^{TR} \cdot k_{ij}^{EU} \cdot G_i \cdot \frac{V_{ij}}{L_{ij} \cdot 2} \cdot K^{(I)} \tag{2.1}$$

or

$$W_{ij} = k_i^{TR} \cdot k_{ij}^{EU} \cdot H_{ij} \cdot K^{(II)} \tag{2.2}$$

where:

k_i^{TR} – technical availability coefficient of the i-th truck, $k_i^{TR} \in [0;1]$ [share units]; technical availability coefficient of the planning period can be determined from the results of the enterprise work for the period prior to the planned one;

k_{ij}^{EU} – fleet utilization coefficient is the degree of utilization of the i-th truck during the work with the j-th excavator for the planning period. Here it is determined as a percentage of k_i^{TR} : $k_{ij}^{EU} = D_i^{TU} / D_{ij}^{P}$, where D_i^{TU} – the amount of days of the i-th truck being in the operative condition, D_{ij}^{P} – the amount of operation days of the i-th truck working with the j-th excavator; the fleet utilization coefficient for the planned period can be determined from the results of the enterprise work for the period prior to the planned;

G_i – mined rock volume, transported by the *i*-th truck, [m^3];

V_{ij} – average speed of the *i*-th truck, [km/h] (here not only the speed of movement is taken into account, but also the time of loading and unloading the truck when working with the *j*-th excavator, and also the complexity of the road profile and road conditions);

L_{ij} – distance of the mined rock transportation by the *i*-th truck from the *j*-th excavator, [km];

$K^{(I)}$ – conversion factor for the reduction to a single dimension of the planning period, for example: for the monthly planning $K^{(I)}$ = 720 h;

H_{ij} – production rate of the *i*-th truck for transportation of the mined rock when using *j*-th excavator, [m^3/shift];

$K^{(II)}$ – conversion factor for the reduction to a single dimension of the planning period, for example: for the monthly planning $K^{(II)}$ = 60 shifts of 12 hours.

1. k_{ij} – indication of using the *i*-th truck during the work with the *j*-th excavator, can be:

- $k_{ij} = 0$, if the *i*-th truck does NOT work with the *j*-th excavator;

- $k_{ij} = 1$, if the i-th truck operates with the j-th excavator.
- Besides, the following restrictions apply for k_{ij}:
- with any excavator more than one truck can operate:
-

$$\sum_{j=1}^{M} k_{ij} \geq 1,$$

- any truck can be operated with only one excavator:
-

$$\sum_{i=1}^{N} k_{ij} \leq 1,$$

- choosing the initial values k_{ij} the conditions of compliance of certain truck models for the work with specific models of excavators must be taken into account (these ratios are set constants and are not changed further for for the solution of the optimization problem).

Also non-negativity condition of variables should be kept, that is:

$$W_j^* > 0;\ W_{ij} > 0;\ G_i > 0;\ V_{ij} > 0;\ L_{ij} > 0.$$

When using this optimization criterion (1) in the MES-systems [4] it is convenient to use formula (2.1), which explicitly involves the average speed of trucks movement. Information about the average speed can be obtained automatically from satellite vehicle tracking systems used on the open-cast mines.

Formula (2.2) is used in the case if it is impossible to obtain information on the actual average trucks speed, and at the mining enterprise the norms for transportation of the mined rock are used.

The presented optimization criterion (1) does not take into account the types of material to be transported. If the output norms for transportation of different types of the mined rock (coal, rock piles and diluted run rock mass) are significantly different from each other, then the formula (1) is represented as follows:

$$\Delta W = \sum_{j=1}^{M} \left| \sum_{b=1}^{B} W_{bj}^* - \sum_{i=1}^{N} k_{ij} \cdot W_{ij} \right| \rightarrow \min \qquad (1.1)$$

where: W_{bj}^* – the planned amount of work for a month for the j-th excavator with b-th rock mass type, j – type excavator, $j \in M$, b – rock mass type, $b \in B$, [m^3/month]. In this case formula (2.2) also can be represented as follows:

$$W_{ij} = k_i^{TR} \cdot k_{ij}^{EU} \cdot \sum_{b=1}^{B} \left(H_{bij} \cdot K_b^{(III)} \right) \qquad (2.3)$$

where: H_{bij} – the *i*-th truck output norm for transportation of the *b*-th mined rock type when using *j*-th excavator, [m^3/shift]; $K_b^{(III)}$ – the estimated number of shifts for transportation of the *b*-th mined rock type.

Furthermore, this problem, in general, is open, since the total performance does not match the excavator trucks performance:

$$\sum_{j=1}^{M} W_j^* \neq \sum_{j=1}^{M}\sum_{i=1}^{N} k_{ij} \cdot W_{ij}.$$

Thus, it is necessary to make it a closed problem. If the trucks production capacity is higher, then it can be done either by introducing a fictitious excavator (in this case, unclaimed trucks would not be used in the process), or by changing the trucks operating conditions (for example, if the open-cast mine has several dumps then some trucks can be assigned to transport the rock to the remote dumps, i.e. L_{ij} increases)). If the excavators production capacity is higher, then the fictitious trucks are introduced into the problem condition (in other words, the question of attracting additional production capacity during the planning period is solved), or the problem of organization concerning the excavator is solved (e.g. at the excavator is a subject of scheduled preventive or other repair works).

Before making the calculations it is necessary to determine the fundamental ability to perform the job by the excavators available. You can use the formula for determining shift productivity of the excavator [2]:

$$W_{bj} = \frac{3600 \cdot G_j^K \cdot k_{bj}^E \cdot k_j^{EU}}{T_{bj}^C} \cdot T^{SH} \tag{3}$$

where: W_{bj} – shift productivity of the *j*-th excavator with the *b*-th mined rock type, *j* – type excavator, $j \in M$, b – mined rock type $b \in B$ [m^3/shift];

G_j^K – bucket capacity of the *j*-th excavator, [m^3];

k_{bj}^E – excavation rate of the *b*-th mined rock type [share units];

k_j^{EU} – utilization coefficient of the *j*-th excavator during the shift, [share units];

T_{bj}^C – duration of the excavation cycle of *b*-th mined rock type [s];

T^{SH} – shifts duration, [hour].

Comparison of W_{bj} and W_{bj}^* will show either a fundamental opportunity to perform excavation plan $\left(\sum_{j=1}^{M} W_{bj} \geq \sum_{j=1}^{M} W_{bj}^*\right)$, or inability to perform with the existing equipment. $\left(\sum_{j=1}^{M} W_{bj} < \sum_{j=1}^{M} W_{bj}^*\right)$.

The functional structure of the model is shown in the Figure 1.

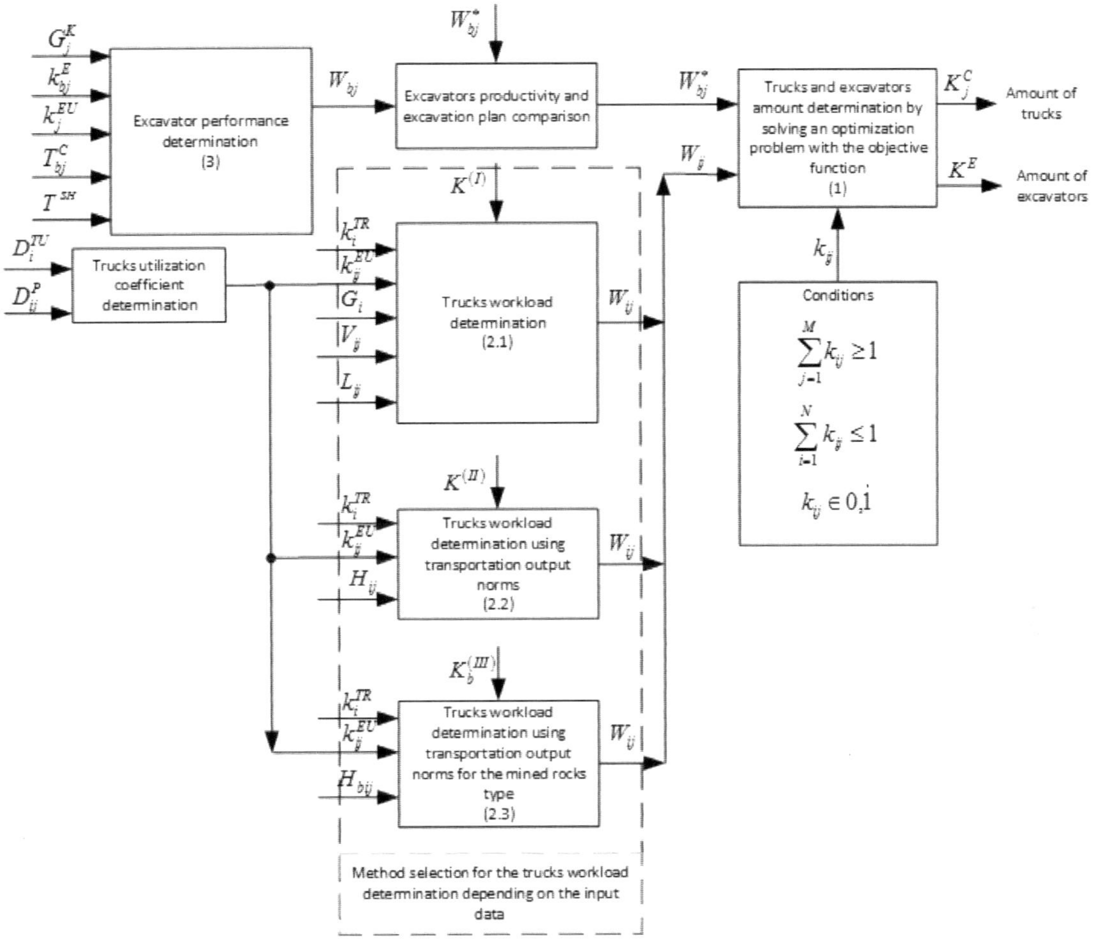

Figure 1. The functional structure of the model.

By varying signs of equipment utilization k_{ij} the minimization of the objective function ΔW can be achieved. For example, the potential method can be used for this purpose, which is a modification of the simplex method for solving linear programming problems with regard to the transport problem.

3. An example of solving the problem of trucks assignment

Consider the example of solving the problem. Suppose two excavators operate at the open-cast mine: EKG-5A (bucket capacity 5 m^3) and Komatsu PC1250 (bucket capacity 6 m^3). Suppose according to the terms of mining the excavators are to extract the mined rock in the amount of 700 thousand m^3 (200 thousand m^3 and 500 thousand m^3, respectively) a month. The motor depot of the mining enterprise has 10 trucks: 3 trucks BelAZ 7555B (body capacity 22 m^3), 3 trucks Komatsu HD 785-7 (body capacity 60 m^3), 4 trucks Volvo A35F (body capacity 20 m^3). It is necessary to assign trucks if a unit have the following characteristics (Table 1). For simplicity of calculations we specify the performance characteristics of the trucks and their makes.

Table 1. Performance characteristics of the trucks.

Characteristics	BelAZ 7555B		Komatsu HD 785-7		VolvoA35F	
	EKG-5A	PC1250	EKG-5A	PC1250	EKG-5A	PC1250

k_{ij}^{EU}	0.9	0.8	1	1	1	0.9
k_i^{TR}	0.75		1		0.5	
G_i	22		60		20	
V_{ij}	14	15	9	10	14	15
L_{ij}	2	3	2	3	2	3

The calculations made with the help of Microsoft Excel tool "Search solution" (Table 2) showed that the deviation from the plan is less than 1%. For the plan implementation it is sufficient to use 8 trucks. The remaining trucks can be left in reserve, for example, or to recalculate the problem by increasing the transportation distance of the mined rock.

Table 2. Calculation results.

	k_{ij}	Excavators		Result
		EKG-5A	Komatsu PC-1250	
Trucks	BelAZ 7555B No. 1	1	0	Assigned
	BelAZ 7555B No. 2	0	0	Not assigned
	BelAZ 7555B No. 3	1	0	Assigned
	Komatsu HD 785-7 No. 1	0	1	Assigned
	Komatsu HD 785-7 No. 2	0	1	Assigned
	Komatsu HD 785-7 No. 3	0	1	Assigned
	VolvoA35F No. 1	0	1	Assigned
	VolvoA35F No. 2	0	1	Assigned
	VolvoA35F No. 3	0	0	Not assigned
	VolvoA35F No. 4	1	0	Assigned
Execution plan difference		-88 m^3	3200 m^3	

4. Verification of solution feasibility

The practical application of the proposed method of trucks assignment may be accompanied by verification of the feasibility of the solution found. It is more effective to use a two-level check of the technical capacity of production plans execution and possibility for introduction of the found trucks assignment.

Verification of the technical capacity to meet the production plans is carried out before the calculations and consists in the determination of the excavators productivity by formula (3).

The correctness of the found solutions can be verified by the statistical simulation model of the production process, built on the basis of queuing systems [1]. It is also possible to determine the service rate of trucks by each every excavator, relative and absolute capacity of excavators, their idle waiting for loading, etc.

Thus, the trucks service rate by the j-th excavator can be determined as follows:

$$\nu_j = med\left(\frac{60}{T^C_{bj} \cdot \dfrac{G_j}{k_{bj} \cdot G^K_j}}\right) \tag{4}$$

where: T^C_{bj} – time of the excavation cycle for the b-th type of the rock mined by the j-th excavator, [s];

G_i – body volume of the i-th truck, [m^3];

G^K_j – bucket volume of the j-th excavator, [m^3];

k_{bj} – utilization rate of the j-th excavator bucket capacity when excavating the b-th type of the mined rock.

Trucks flow rate can be determined either by using output norms:

$$\lambda_j = \frac{1}{580} \cdot \sum_{i=1}^{K^C_j} H^X_{bij} \tag{5.1}$$

where: H^X_{bij} – number of hauls of the i-th truck per shift during transportation of the b-th type of the mined rock when using the j-th excavator, [pc/min];

K^C_j – number of trucks assigned to work with the j-th excavator.

Or by using the initial data on the vehicle operation:

$$\lambda_j = med\left(\frac{1}{\dfrac{L_{ij} \cdot 2}{V_{ij}} \cdot 60 + T^D_i}\right) \tag{5.2}$$

where: T^D_i – i-th truck unloading time, [min].

Relative capacity of the j-th excavator is defined by the formula:

$$q_j = \frac{\nu_j}{\lambda_j + \nu_j} \tag{6}$$

If q_j is less than 1, then the excavator will not have time to serve all coming for loading trucks – the trucks queue will be created causing the trucks downtime. If q_j is more than 1, the excavator will stand idle, waiting for the truck to be loaded. According to the calculated capacity of the excavator it is necessary to make organizational conclusions and decide whether to recalculate the trucks assignment with the modified initial conditions, or to adopt the obtained equipment assignment.

5. Conclusions

Use of the proposed method makes it possible to automate the process of trucks distribution between the excavators taking into account the optimization criterion of production plans fulfillment. Check of the specified optimization criterion on the model example successfully demonstrated the possibility of its practical application. As a result of the simulation the transport assignment was performed, allowing the technological transportation with minimal deviation from the production plan to be performed.

At the moment the implementation of the proposed method of vehicle assignment with the help of the software product "1C: ERP Mining Enterprise 2" is carried out, designed for automation of operational, managerial and regulated accounting and planning of mining enterprises.

References

[1] Reznichenko S S 1991 *Economic-mathematical Methods and Modeling in the Planning and Management of Mining Production* (M.: Nedra) p 429
[2] Sysoev A A 2005 *Engineering and Economic Calculations for the Opencast Mining* (Kemerovo: KuzSTU) p 179
[3] Kuharev V N 1991 *Economic-mathematical Methods and Models in Planning and Management* (Kiev: Vyshchashk) p 304
[4] Andreev E B 2015 *MES-systems: A View From the Inside* (M.: RTSoft) p 240

Modelling of CWS combustion process

I A Rybenko and L A Ermakova

Siberian State Industrial University, 42 Kirova Street, Novokuznetsk, Russia, 654007

E-mail: ermakova@sibsiu.ru

Abstract. The paper considers the combustion process of coal water slurry (CWS) drops. The physico-chemical process scheme consisting of several independent parallel-sequential stages is offered. This scheme of drops combustion process is proved by the particle size distribution test and research stereomicroscopic analysis of combustion products. The results of mathematical modelling and optimization of stationary regimes of CWS combustion are provided. During modeling the problem of defining possible equilibrium composition of products, which can be obtained as a result of CWS combustion processes at different temperatures, is solved.

1. Introduction

In recent years, requirements to the quality of the energy supplies used in industry and in the housing and utilities infrastructure have significantly increased, which in turn led to the development of coal processing technology. As a result of the processing the fine wastes (fine particles of coal and mineral inclusions) are formed, which are placed into the piles polluting the environment.

An effective method of disposing waste coal is its combustion in the vortex adiabatic chamber in the form of CWS [1-3]. This approach to utilization allows the mined coal to be used for generation of heat and energy, and the ash produced is a good building material.

To develop CWS combustion technology in the vortex adiabatic chamber it is necessary to study the process of CWS drops combustion, as well as to solve the problem of modelling and optimization of static combustion modes of CWS in the vortex chamber.

2. Physico-chemical scheme of CWS drops combustion

Describing the process of CWS combustion it is necessary to consider the main differences between the process of ignition and burning of a slurry drop from the combustion of pulverized solid fuel. As the laboratory studies presented in [4-9] showed CWS drops combustion occurs simultaneously with the evaporation of moisture from its inner part, and evaporation of moisture does not slow down the carbon combustion process. The released vapor jets from the inner part of the drop activate the combustion on the drop surface, as a result of the simultaneous flow of these two processes, the suspension drop turns into a highly porous body [4]. To describe the combustion process of CWS drops the following physico-chemical process scheme consisting of several independent parallel-sequential stages can be offered:

- *on the drop surface* – instant heating and surface evaporation, low temperature activation of the fuel reaction surface prior to its ignition, combustion of coal microparticles on the drop surface;

- *inside the drop* – gradual drop heating, evaporation of moisture from the inner part of the drop, volatile yield and combustion of volatiles near CWS drop, transfer of coal microparticles to the drop surface with the formation of a porous cenospheres (hollow spheres), combustion of coal microparticles as a result of chemical reaction with oxygen and vapour.

When the CWS drop gets into the hot combustion chamber its surface almost instantaneous gets heated and the moisture evaporation heating of the surface of the drops and the moisture evaporation from the drop surface. It is followed by the moisture evaporation from the inner part of the drop. The process of evaporation is described by the equation:

$$H_2O = \{H_2O\}. \quad (1)$$

As the temperature of the drop surface rises the pyrolysis of organic matter on the CWS drop surface begins, accompanied by the yield of volatile substances. With the temperature front advancing inside the drop the process of volatiles yield from the inner part of the drop starts. The basic equations describing the volatile yield are as follows:

$$\begin{aligned} CO_2 &= \{CO_2\}, \\ H_2 &= \{H_2\}, \\ CH_4 &= \{CH_4\}, \\ CO &= \{CO\}. \end{aligned} \quad (2)$$

The emitted flammable gaseous components react with oxygen [4]:

$$\begin{aligned} &2\{H_2\} + \{O_2\} = 2\{H_2O\}, \\ &\{CH_4\} + 2\{O_2\} = \{CO_2\} + 2\{H_2O\}, \\ &2\{CO\} + \{O_2\} = 2\{CO_2\}. \end{aligned} \quad (3)$$

Due to the water vapour, formed during the moisture evaporation from the surface, the low temperature activation of coal microparticles on the CWF drop surface occurs leading to their ignition, which results in a significant reduction in the CWS ignition temperature. In the process of volatiles combustion the heating of drops takes place and the coke residual of coal microparticles on its surface ignites. On the surface the combustion reactions take place due to the interaction of carbon with oxygen, water vapour and CO_2, obtained because of the volatiles yield, combustion of volatiles and coke base of fuel. Combustion reactions occur in accordance with the following reactions [4]:

$$\begin{aligned} &C + \{O_2\} = \{CO_2\}, \\ &2C + \{O_2\} = 2\{CO\}, \\ &C + \{CO_2\} = 2\{CO\}, \\ &C + \{H_2O\} = \{CO\} + \{H_2\}. \end{aligned} \quad (4)$$

The combustion process on the surface of CWS drop takes place in parallel with the processes of evaporation and volatiles yield from the depth of the drop. During the process of moisture evaporation and volatiles yield the coal microparticles move from the inner part of the drop to its surface [4], at the same time, water evaporation zone expands to the inner layers of a drop, thus, causing an increased pressure and enlargement of the drop size [4]. On the drop surface the carbon particles burn out, and the mineral part cokes forming a porous hollow sphere. Graphic scheme of the combustion stages of CWS drops is given in [10].

3. Particle size analysis of CWS and its combustion products

To conduct the study a laser analyzer of particle sizes Mastersiser 2000 (Malvern, UK) was used. Table 1 shows the results of experimental study on particle size distribution of CWS produced from the waste coal at processing plant of mine "Antonovskaya" and two dust samples taken in the vortex dust collector in the course of the experiment on CWS combustion in the vortex adiabatic chamber.

Table 1. Particle size distributions of CWS and ash.

	Range of sizes, mm	0.00002-0.00025	0.00025-0.00071	0.00071-0.00200	0.00200-0.00564	0.00564-0.01589	0.01589-0.04478	0.04478-0.12619	0.12619-0.35566	0.35566-2.000
Amount of particles, %	CWS	0	0.65	2.25	7.36	20.35	33.65	29.16	6.47	0
	Ash (sample No.1)	0	0	0	0.11	6.56	43.02	41.32	8.31	0.7
	Ash (sample No.2)	0	0	0	1.21	8.95	20.55	39.06	28.18	2.06

The estimate of the density distribution *f(d)* of particle by sizes on the basis of experimental data is performed in the class of regular distributions in the range (d_{min}, d_{max}) in the form of a lognormal distribution taking into account the minimum and maximum particle diameters d_{min} and d_{max}:

$$f(d)=\frac{\lambda(d_{\max}-d_{\min})}{(d_{\max}-d)(d-d_{\min})\sigma\sqrt{2\pi}}e^{-\frac{\left(\ln\left(\frac{d_{\max}-d}{d-d_{\min}}\right)-\ln a\right)^2}{2\sigma^2}}, \quad (5)$$

where λ, a – distribution parameters.

The specific form $\tilde{f}_m(d)$ is determined by the values of parameters λ*, a* and □*, which minimize the functional:

$$\Phi(\lambda^*,a^*,\sigma^*)=\min_{\lambda,a,\sigma}\sum_{i=1}^{N_f}(f(d_i)-S_i)^2, \quad (6)$$

where N_f – number of fractions of the corresponding substance;

d_i – particle size in the i-th fraction, mm;

S_i – share of the i-th fraction%.

The obtained values of particle density distribution by sizes for CWS and products of its combustion are given in Table 2.

Table 2. Values of particle density distribution by sizes for CWS and ash.

Material	d_{min}, mm	d_{max}, mm	λ	a^*	σ^*
CWS	0.00028	0.31698	0.611	2.555	1.177
Ash (sample No. 1)	0.00448	0.63246	0.703	7.668	0.846
Ash (sample No. 2)	0.00317	2.00000	1.701	8.194	1.011

Figure 1 shows the experimental data in the form of histograms of CWS and ash particles distribution by sizes and the resulting graph of density distribution in the form of equation (5). The particle sizes are arranged on a logarithmic scale.

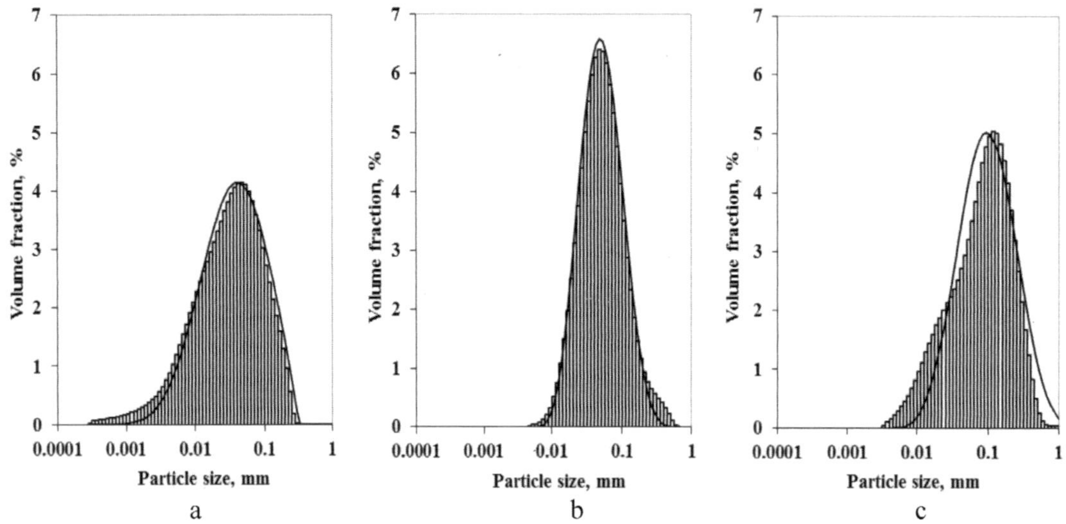

Figure 1. Histograms of CWS and ash particles distribution by sizes: a – CWS; b – ash (sample No. 1); c – ash (sample No. 2).

Research of CWS combustion products. The chemical composition of ash, sampled in the vortex dust collector in the course of the experiment on the CWS combustion in the vortex adiabatic chamber, was defined by X-ray fluorescent analysis. For X-ray fluorescent analysis the scanning (single-channel) spectrometer XRF-1800 (Shimadzu, Japan) was used with the X-ray tube with Rhodium anode, voltage – 40 kV and current strength – 95 mA, calculating time for all elements – 552 seconds, the calculation method – method of fundamental parameters. Table 3 shows the component composition of ash.

For stereomicroscopic study of dust sampled in the vortex dust collector in the course of the experiment on CWS combustion in the vortex adiabatic chamber, the sample was dissolved in water and divided in it into 2 fractions. The heavy fraction 80-85% is located at the bottom of the assay cup, the light one floats to the surface of water. These fractions were separately investigated under the stereomicroscope with magnification from 20 to 300 times. All particles are well visible, of spherical type, or, more often, their fragments (Figure 2).

Table 3. Component composition of the ash,%.

Sample No.	C	SiO_2	Al_2O_3	Fe_2O_3	S	CaO	K_2O	Na_2O	MgO	TiO_2	P
Ash (sample No. 1)	8.05	51	17.23	6.18	0.68	4.61	2.75	2.47	1.99	1.01	0.14
Ash (sample No. 2)	5.47	53.7	18.38	5.23	0.62	4.59	2.85	2.47	2.25	0.94	0.18

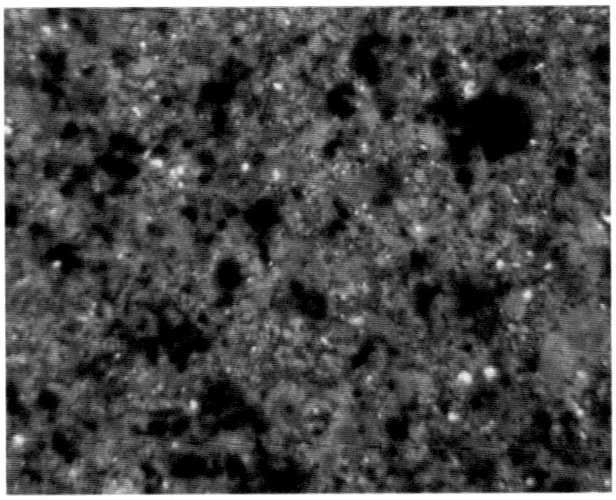

Figure 2. Stereomicroscopic investigation of dust from the cyclone (Reflected light. Magnification 400x).

The light part is composed of graphite particles and other mineral particles. The graphite particles form either whole hermetically closed micro spherulites or segmental fragments of these spherulites. Graphite is recrystallized, but the crystals are very small, except for spherical and hemispherical particles there are linearly stretched graphite particles. They represent an aggregate of plate micaceous crystals. The spherical shape of particles creates an impression about their significant predominance over other particles. The spherical graphite particles are shiny, black, have a proper shape of a sphere or elongated ovaloid, the spheres sizes are different from 2-3 micrometers to 100 micrometers.

The brownish-yellow particles are presented by silicate glass, colored by iron oxides. Large particles are agglomerate of glassy mocrosphaerolites. Medium and small isometric particles are larger mocrosphaerolite; they are more often represented by segment fragments. The silicate glass content in the ash reaches 48%, in the cyclone dust ranges from 3.5 to 35%.

The crystalline silica is represented mainly by crystobalite, more seldom α- and β-quartz; their total content ranges from 0.5 to 8.5%. The carbides and nitrides of silicon, aluminum and alkaline earth metals have a greenish-gray color and are about 1%. Especially there is much hollow graphite mocrosphaerolites in the dust. By volume, they represent in different samples from 5 to 35%, by weight from 2 to 18%. There are particles of an emerald-green color; and in the beginning they could not be seen, they appeared with the course of time. It was found that it is iron-copper sulfate (chalcanthite) due to the presence of calcium, iron and copper sulfides in a sufficient amount in the ash (about 2%).

These results agree well with the data [4] obtained while investigating the ash particles produced by combustion of dry dust and coal-water slurry from the same coal: during the slurry combustion the number of ash particles reduces, their sizes increases (from 0.01-0.05 to 0.5-1.5 mm) and the surface structure changes.

4. Mathematical modelling of stationary regimes of waste coal combustion in the adiabatic chamber

To solve the problems of modelling and optimization of static modes of CWS combustion in the vortex chamber a methodology and calculation system, based on the thermodynamic modeling, calculations of material and heat balances and solving optimization problems, was developed.

Thermodynamic modelling allows various tasks to be solved with the help of software "Terra" [10], which implements the principle of maximum entropy and thermodynamic analysis of co-occurrence in the system of possible chemical reactions on the basis of the calculation of changes in Gibbs energy and equilibrium constants.

At the stage of the calculation of the material and heat balances a flexible choice of a input and output parameters of the system is foreseen, taking into account the results obtained at the stage of thermodynamic modeling for the chosen type of process and the range of parameters; and it makes possible to obtain a mathematical description of the relationship of flows and process parameters, which are the basis for solving the optimization problem concerning the determination of the costs of materials at a selected optimization criteria and constraints.

In the thermodynamic modelling the process of CWS combustion in the vortex chamber was regarded as a complex thermodynamic system, which receives condensed (water coal fuel) and gaseous (air) incoming flows; the final products are ash and gas. With the flows the following substances get into the reactor: condensed phase – *C, O, H, N, P, As, HCl*, SiO_2, Al_2O_3, Fe_2O_3, *CaO*, *MgO, MnO*, TiO_2, P_2O_5, *S*, K_2O, Na_2O, H_2O; gas phase – O_2, N_2. The elements constituting these compounds are: *Fe, C, O, As, HCl, Mn, Si, S, P, Al, Ca, Mg, Na, K, H, Ti, N.*

The mass balance equations were developed on the basis of the law of mass conservation with respect to the components of a heterogeneous system. The main feature of the calculation is the examination of the process in the form of a balanced system on three levels: flows, substances and elements:

$$\sum_{k=1}^{K^f} G_k^f + \sum_{k=1}^{Kg} G_k^g \; \sum_{=l=1}^{L^f} G_l^f + G^g \quad ;$$

$$\sum_{k=1}^{K^f}\sum_{m=1}^{N_k^f} \frac{G_k^f / R_m /_k}{100} + \sum_{k=1}^{K^g}\sum_{m=1}^{N_k^g} \frac{G_k^g \{R_m\}_k}{100} = \sum_{l=1}^{L^f}\sum_{n=1}^{N_l^f} \frac{G_l^f / R_n /_l}{100} + \sum_{n=1}^{N_g} \frac{G_g \{R_n\}}{100};$$

$$\sum_{k=1}^{K^f}\sum_{m=1}^{N_k^{E_i^f}} \frac{G_k^f / E_{ix_m} E_{jy_m} /_k}{100} \cdot \frac{x_m M_{E_i}}{M_{E_{ix_m}E_{jy_m}}} + \sum_{k=1}^{K^g}\sum_{m=1}^{N_k^{E_i^g}} \frac{G_k^g \{E_{ix_m} E_{jy_m}\}_k}{100} \cdot \frac{x_m M_{E_i}}{M_{E_{ix_m}E_{jy_m}}} =$$

$$= \sum_{l=1}^{L^f}\sum_{n=1}^{N_l^{E_i^f}} \frac{G_l^f / E_{ix_n} E_{jy_n} /_l}{100} - \frac{x_n M_{E_i}}{M_{E_{ix_n}E_{jy_n}}} + \sum_{n=1}^{N_l^{E_i^g}} \frac{G_l^g \{E_{ix_n} E_{jy_n}\}_l}{100} \cdot \frac{x_n M_{E_i}}{M_{E_{ix_n}E_{jy_n}}}, \quad (7)$$

where K^f, K^g – number of condensed and gaseous flows;

G^f_k, G^g_k – mass consumption of condensed and gaseous flows, kg/s;

L^f – number of condensed output flows;

G^f_l, G^g – mass yield of the *l* condensed and gaseous flows, kg/s;

N^f_κ, N^g_κ – amount of substances in the *k* flow;

R_{mk}, $\{R_m\}_k$ – substance content R_m in the *k* condensed and gaseous input flow, %.

$/E_{ix_m}E_{jy_m}/_k$, $\{E_{ix_m}E_{jy_m}\}_k$ – concentration of *m* substance containing element E_i in the condensed *k* input gaseous flow, %;

$/E_{ix_n}E_{jy_n}/_l$, $\{E_{ix_n}E_{jy_n}\}$ – concentration of *n* substance containing element E_i in *l* condensed or gaseous output flow, %;

$N_k^{E_i^f}$, $N_k^{E_i^g}$ – amount of substances containing the element E_i in *k* condensed gaseous input flow, respectively;

$N_l^{E_i^f}$, $N^{E_i^g}$ – amount of substances containing the element E_i in *l* condensed gaseous output flow;

x_m, y_m, x_n, y_n – stoichiometric coefficients of *m* and *n* element compound E_i;

M_{Ei}, $M_{E_{ix_m}E_{jy_m}}$, $M_{E_{ix_n}E_{jy_n}}$ – molar mass of element E_i and its compound, kg/mol.

$$\sum_{k=1}^{K^f}\sum_{m=1}^{N_k^f} (\Delta H_T^0)_m \frac{G_k^f / R_m /_k}{100 \cdot M_{R_m}} + \sum_{k=1}^{K^g}\sum_{m=1}^{N_k^g} (\Delta H_T^0)_m \frac{G_k^g \{R_m\}_k}{100 \cdot M_{R_m}} =$$

$$= \sum_{l=1}^{L^f} G_l^f \Delta H_l^f + G^g \Delta H^g + \sum_{w=1}^{W} \Delta H_w^0 \left(\sum_{k=1}^{K} \frac{G_k R_{kw}}{100} - \sum_{l=1}^{L} \frac{G_l R_{lw}}{100}\right) + Q_{los}. \quad (8)$$

The heat balance equations are constructed on the basis of energy conservation law. The main defining processes for the thermal state are the inflow and outflow of heat through the input and output material flows; the heat exchange with the environment and chemical reactions with the appropriate thermal effects:

$$\sum_{k=1}^{K^f} G_k^f \Delta H_k^f + \sum_{k=1}^{K^g} G_k^g \Delta H_k^g + = \sum_{l=1}^{L^f} G_l^f \Delta H_l^f + G^g \Delta H^g + \sum \Delta H_{\text{c.r.}} + Q_{los}, \quad (9)$$

where ΔH_k^f, ΔH_k^g – enthalpy of a unit mass of k condensed or gaseous input flows with respect to the temperature 298 K, kJ/kg;

ΔH_l^f, ΔH^g – enthalpy of a unit mass of l-го condensed or gaseous output flows with respect to the temperature 298 K, kJ/kg;

Q_{los} – heat losses into to the environment, kJ;

$\sum \Delta H_{\text{c.r.}}$ – total heat effect of chemical reactions, kJ.

Using the developed methodology and calculation system the calculations and studies of the CWS combustion process in the vortex chamber were performed.

The initial data for calculation were as follows: the calculation is made for the CWS consumption 1 tonne/h, moisture – 37%, ash content – 37.5%. The fuel combustion is carried out with the use of air. The chemical composition of the organic part of the sludge and are ash shown in Table 1.

Table 4. CWS composition.

Organic part, %		*C*	*H*	*N*	*O*	*P*	*As*	*Cl*			
		82.588	5.50	2.50	9.30	0.041	0.004	0.068			
Ash, %	SiO_2	Al_2O_3	Fe_2O_3	CaO	MgO	TiO_2	MnO	P_2O_5	*S*	Na_2O	K_2O
	60.455	19.00	4.20	7.25	2.00	0.725	0.02	1.05	2.75	0.55	2.0

Tables 2 and 3 contain the calculated heat and material balances for CWS consumption 75 l/h (90 kg/h). The flue gas temperature at the given parameters is 1345 °C.

Table 5. The material balance of CWS combustion.

Input flows	Consumption, kg/h	Output flows	Yield, kg/h
CWS	90.000	Gas	476.187
Air	406.927	Ash	20.740
Total:	496.927	Total:	496.927

Table 6. The heat balance of CWS combustion.

Input	kg/h	MJ/kg	MJ/h	Consumption	kg/h	MJ/kg	MJ/h
Exothermic reactions			1122.4	Products enthalpy			827.3
$C+1/2O_2=CO$	2.92	9.6	28.0	Ash	20.7	1.6	33.1
$C+O_2=CO_2$	26.3	33.1	873.6	Gas	476	1.7	794.2
$H_2+1/2O_2=H_2O$	1.75	125.9	220.8	Endothermic reactions			182.9
				Moisture evaporation	33.3	5.5	182.9
				Heat losses			112.0
				Residual error			0.00%
TOTAL INPUT			1122.4	TOTAL CONSUMPTION			1122.4

Using the developed methods the combustion modes of CWS, produced from waste coal, were studied at various moisture values, ash content (Figure 3), CWS and air consumption rates.

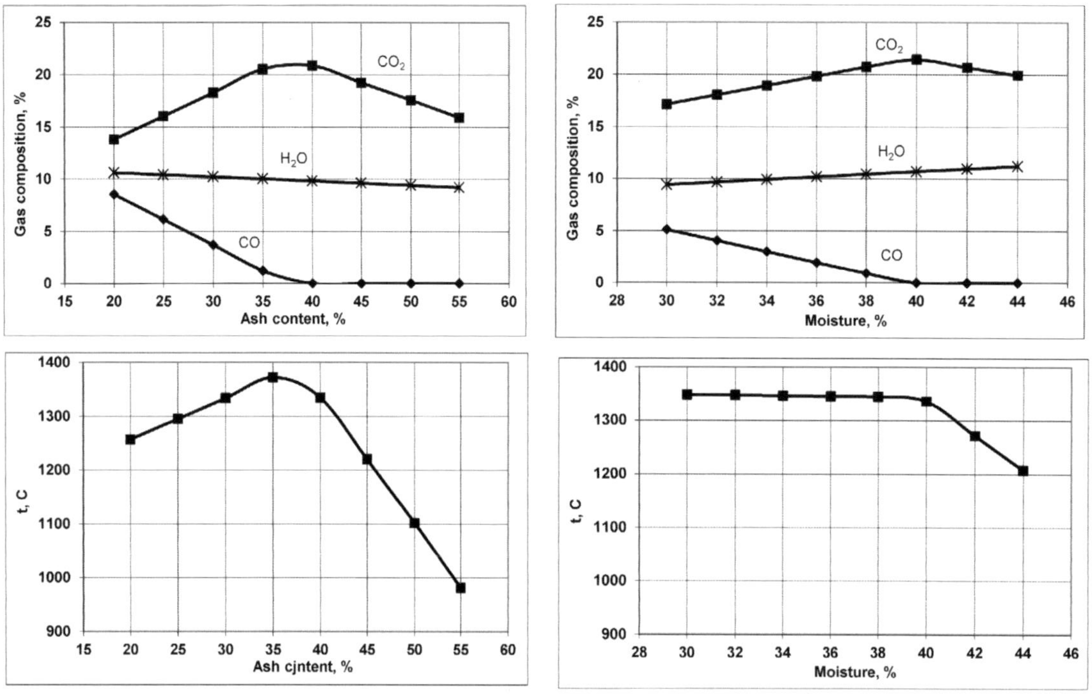

a b

Figure 3. Dependence of the parameters of CWS combustion on its composition (a – on ash, b – on moisture).

This method can be used to calculate the stationary CWS combustion modes of any composition and ti solve the tasks of defining the necessary basic technological modes.

5. Conclusions

Thus, the conducted particle size distribution analysis and stereomicroscopic study confirm that as a result of CWS combustion hollow spherical particles of larger sizes than the initial CWS are formed, which corresponds to the proposed CWS combustion process description above.

The mathematical modelling of stationary regimes of CWS combustion in the vortex chamber is carried out. The method of calculation of stationary CWS combustion regimes in the vortex chamber is developed, which is based on calculations of material and heat balances and solution of the optimization problem. With the use of the developed method and calculation system the the stationary modes of CWS combustion in the vortex chamber are calculated, the combustion process parameters are defined for various CWS characteristics (moisture, ash content). The mathematical modelling of CWS combustion at different consumption rates of CWS and air is performed.

References

[1] Sarychev V D, Kuksov I A et al 2012 *Steel in Translation* B **42** (8) 620–623
[2] Kuksov I A, Mochalov S P, Sarychev V D 2012 *World Applied Sciences Journal B* **19** (1) 100–105
[3] Mochalov S P, Rybenko I A et al 2012 *Steel in Translation B* **42** (8) 611–613
[4] Kantorovich B V and Delyagin G N 1962 *Heat Mass Transfer* **5** 11
[5] Kobayashi H, Howard J B and Sarofim A F 1976 *Proc. of the 18th Symp. on Combustion* (Pittsburgh: Combustion Institute) p 411

IOP Conf. Series: Earth and Environmental Science **45** (2016) 012016 doi:10.1088/1755-1315/45/1/012016

[6] Smoot D J and Smith P J 1985 *Coal Combustion and Gasification. The Plenum Chemical Engineering series, New York*
[7] Kijo-Kleczkowska A 2011 *Archives of Thermodynamics* **32** (1) 45–75
[8] Dunn-Rankin D, Hoornstra J, Gruelich F A and Holve J 1987 *Fuel* **66** (8) 1139–45
[9] Murdoch P L, Pourkashanian M and William A 1985 *Proc. of Symp. on Combustion* **20** (1) 1409–18
[10] Mochalov S P, Rybenko I A and Ermakova L A 2012 *World Applied Sciences Journal B* **19** (1) 20–25
[11] Sinyarev G V et al 1982 *Computer Application of Thermodynamic Calculations of Metallurgical Processes* (M.: Nauka) p 263

Rational preparation of waste coal mixture for production of bricks by the method of compression molding

A Yu Stolboushkin, A I Ivanov, M V Temlyantsev and O A Fomina

Siberian State Industrial University, 42 Kirova Street, Novokuznetsk, Russia, 654007

E-mail: ivanovaliv1989@gmail.com

Abstract. Rational preparation of the mixture containing technogenic raw material – waste coal for the production of wall ceramics is developed. It was established that the technology of high-quality ceramic bricks requires: grinding of raw materials to class 0.3 + 0 mm, its aggregation in the intensive mixers into granules 1-3 mm, compression molding of adobe to plastic deformation of granules, drying and firing.

1. Introduction

For the modern construction industry, having problems with sustainablesupplyofraw materials of good quality, one of the main directions of development is the utilization of technogenic wastes in the process of construction and technological production [1,2]. As a rule, the stripping ratio – 4 tonne of overburden per 1 tonne of coal causes a considerable pollution of the environment. Obsolete equipment and technology at the majority domestic coal preparation plants lead to a poor quality of coal preparation, as a result, the coal content in the waste ranges from 17 to 26%. This was the main reason for laying-off the brick factory, which more than 20 years used waste coal from the coal processing plant "Abashevskaya" in Novokuznetsk [3].

At the same time, there are examples of innovative technological solutions in our country, which allow us to consider the wastes generated during coal mining and processing not only as a source of raw materials, but also as a power base of enterprises manufacturing ceramic wall materials [4].

The authors carried out research into the problem of waste coal utilization for bricks manufacturing [5], and offered a scheme of provision brick factories with technological fuel obtained after secondary processing of coal (Figure 1).

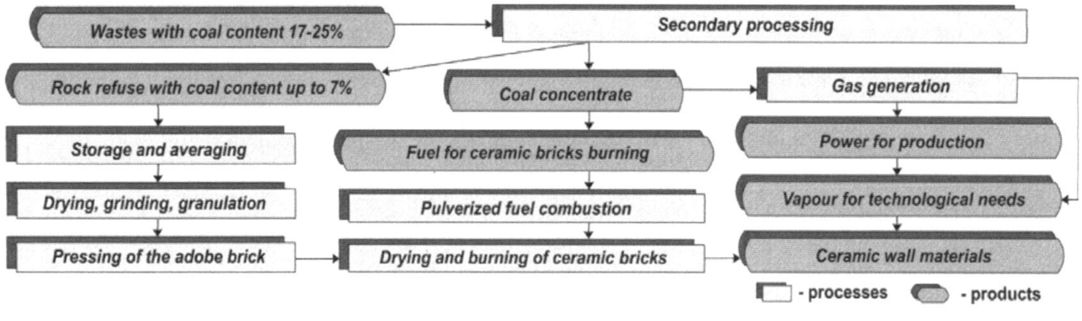

Figure 1. Flow diagram of wall ceramics production from waste coal.

Theproposed technological idea consists in organization of the deep processing of coal right at the brick factory, where the produced fuel will be used for the main production.The "tailings" of stable composition with minimum carbon content formed in the process of enrichment should be used as the main raw material for the production of ceramic brick.

As part of the program on import substitution [6] for the industry of ceramic wall materials the development of domestic technology and equipment, focused on high quality products, with regard to the characteristics of technogenic ceramic raw material from waste coal, is of high importance.

The aim of this work was to develop rational methods for preparation of waste coal mixture for production of ceramic bricks of good quality by compression molding.

2. Rational preparation of raw material

Performance experience gained by the brick factories with the technology of semi-dry pressing showed low quality of products in terms of bending strength and frost resistance compared with the products of plastic molding. It was due to the insufficient grinding of raw materials (up to 3 mm). Heterogeneity of the press powder composition obtained from coarse-grained masses results in a low strength of adobe. The developed method of formation of optimal ceramic brick structures allows authors to avoid structural defects by fine dry grinding, granulation, compression molding, drying and firing [7].

Preparation of the activated fine grained press powders from raw material can be carried out by using modern mixers-granulators of intensive action. Due to the variation in the rotation speed of the blades the regulation of press-powder granulometry within wide limits is possible, and complex turbulent motion of the material ensures active mixing of the adobe components and uniform moisture distribution in the volume of the resulting granulate.

The rational size distribution is a decisive prerequisite for a good compaction and the compact packaging of granules largely determines the final average density of a semi-finished product and purposefully influences on its important properties such as strength, water absorption, frost resistance, etc. Research carried out with the use of traditional technology of semi-dry pressing technique showed that the adobe brick, containing 40-50% of fraction 3-1 mm, 10-20% of fraction 1-0.5 mm and 50-40% of fraction <0.5 mm, has the minimum porosity [8, 9]. We found [10] that the structure of adobe-brick from granulated fine masses with a predominant granule size 1-3 mm is significantly different from the adobe brick of semi dry pressing, produced by the traditional drying-grinding technology of mixture preparation, and gives results that are similar to the plastic molding.

3. Methods and materials

During planning and researching we developed the models of formation of adobe brick structures from granulated press powder (Figure 2). At the initial stage of compression molding of the material there is a reorganization and convergence of granules. Further, as the pressing powder increases, their plastic irreversible deformation with simultaneous air removal from the space between them takes place. It is found that at the optimal compression pressure during adobe compaction to plastic deformation of granules in the surface layer and at the interface between them the concentration of liquid phase occurs due to the squeezing of water from the central part of the granules body (Figure 2c).

The implementation of the described principles of rational mixture preparation for face wall ceramics of compression molding was carried out on the example of waste coal. We used waste coal with class + 13 mm from coal processing plant "Abashevskaya" in Kemerovo region, Russia after extraction of the carbon-containing part (about 7% of the concentrate with Q=4500 kcal/ kg). As a corrective additive lean loam raw materials (Novokuznetsk) was used.

Novokuznetsk loam is the most common type of clay rocks in Siberia. The raw material is low-disperse, non-caking, highly sensitive to drying, mineral composition refers to polymineral rocks of hydromica-kaolinite-montmorillonite type. Waste coal are similar to clays [11] and consist of mineral part represented by hydromuscovite, quartz, plagioclase, montmorillonite, chlorite, siderite, calcite, and carbon residue containing free carbon in the amount from 6 to 8% after the secondary processing.

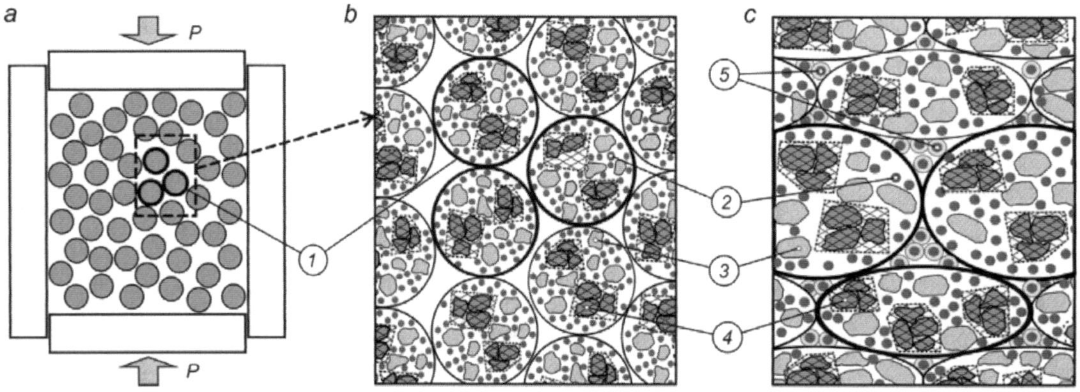

Figure 2. Scheme of compression molding of adobe brick with two-sided application of loading (a), granulated press-powder to plastic deformation of granules (b), structure of the molded adobe brick (c): 1 – granules of ceramic molding material; 2 – clay minerals; 3 – non-clay minerals; 4 – clusters of mineral phase; 5 – colloidal fraction of clay minerals.

According to the chemical composition (Table 1) both types of waste coal belong to the group of semi-acid raw material with a high content of coloring oxides.

Table 1. Chemical composition of waste coal (after processing) and corrective additives.

Raw material	Mass fraction of components to the dried substance, %								
	SiO_2	Al_2O_3	TiO_2	Fe_2O_3	CaO	MgO	K_2O	Na_2O	LOI
Waste coal	57-64	14-17	0.5-1	5-6	3-5	1.5-2	2-2.5	1.5-2	6-8
Loam	61-64	14-15	0.8	4-5	4-4.5	2-2.5	3.5-4		5-6

Drying and milling of raw materials were carried out in a vortex mill-dryer USP-S-04.55M to class -0.3+0. Evaluation of a particle size distribution after grinding of the material was conducted on a laser granulometer Malvern Mastersizer 2000, the results are shown in Table 2.

Table 2. Particle size distribution of waste coal and corrective additives.

Raw material	Fractions content in %, particle size in mm					
	0.315-0.1	0.1-0.05	0.05-0.01	0.01-0.005	0.005-0.001	<0.001
Waste coal	0.02	1.16	25.11	19.80	21.41	32.50
Loam	16.37	36.25	34.81	3.2	4.73	4.64

By particle size distribution waste coal refer to coarse dispersed raw material with a predominance of sand fraction; by ductility – low plastic (plasticity ≈ 6.0); by drying properties – insensitive to drying; by the degree of caking – non-caking material.

Granulation of fine powders was carried out in the intensive mixer to form granules of diameter 1-3 mm with simultaneous moisturisation to the molding moisture. In the process, we determined the optimal parameters for obtaining the granulated mixtures.

In order to improve the product sintering ensured by the formation of the ceramic matrix structure [12], the granulated waste coal was covered with a layer of fine loam. The granulation parameters, material composition and particle size distribution of ceramic mixtures are given in Table 3.

Table 3. Parameters of ceramic mixtures granulation.

No.	Mixture composition	Content, %	W, %	V, s^{-1}	Partial residuals in % on sieves, mm			
					5	2.5	1.2	bottom
1	Loam	100	11.0	125.6	1.5	14.5	72.8	11.2
2	Waste coal Loam	80 20	16.4	157.1	2.5	40.3	33.7	23.5

Laboratory studies showed that the necessary amount of water for powder granulation is 10-17 % [13]. These figures are comparable in moisture values to the extrusion molding technology, therefore, when characterizing the method, generally called "semi-dry pressing", for the developed method of brick production from granulated mixtures it will be more correct to use the term "compression molding".

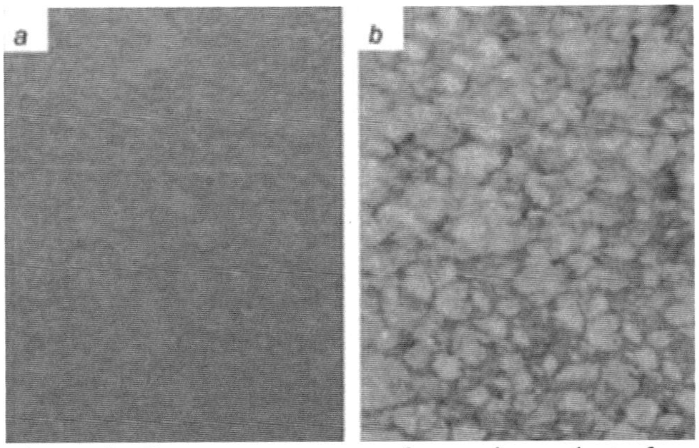

Figure 3. The macrostructure of ceramic products from granulated mixtures of the following composition (%): loam – 100 (a); waste coal – 80 + loam – 20 (b)

The optimization of ceramic mixtures compositions from granulated dispersed powders on the basis of waste coals was performed on samples of 50 mm in diameter and height 45-50 mm, the pressing pressure 12-15 MPa, with a two-sided load application. The samples were fired in the muffle furnace at a maximum temperature 1000-1050 °C. The macrostructure of the obtained ceramic wall materials and the test results of their physical and mechanical properties are presented in Figure 3 and Table 4.

Table 4. Physico-mechanical properties of ceramic samples.

Mixture No.	Compressive strength, MPa	Average density, kg/m^3	Water absorption,%	Frost resistance, cycles	Strength-density ratio
1	19.3	1670	11.6	>50	11.5
2	18.7	1680	18.5	50	16.4

4. Results and discussion

The study on phase composition and microstructure of ceramic samples from waste coal was carried by the complex of methods that includes X-ray phase analysis, petrography of thin sections, SEM, etc. The structure of the ceramic cork produced from waste coal is fine grained, in the granules and on their borders there are non-transparent segregations of brown and dark colour, clearly visible in the parallel and cross polarized light (Figure 4). The burn-out of residual carbon in the granules reacting with oxygen, results in the chemical reactions of reduction inside the material. As a result, during firing in the temperature range 900-950 °C the melt is formed on the boundary and in the central part of the granules.

In the boundary layer the macropores of irregular form elongated along the granules are observed. The disperse phase of the composite material formed from the granules after firing contains conglomerates of fine cryptocrystalline particles sintered together and permeated by micropores. In the body of granules (Figure 4) a significant number of quartz grains of angular shape can be noted. The boundary layer of composite matrix is a cryptocrystalline aggregate continuously transition from one granule into another. It consists mainly of evenly distributed in the in the glass phase of microscopic crystals and fragments of transparent yellowish-white minerals, and fine cryptocrystalline aggregates, predominantly of brownish-red color (Figure 6). On the border of granules there is an intensive formation of glass phase, contributing to the reaction flow and reinforcing binding effect of the matrix.

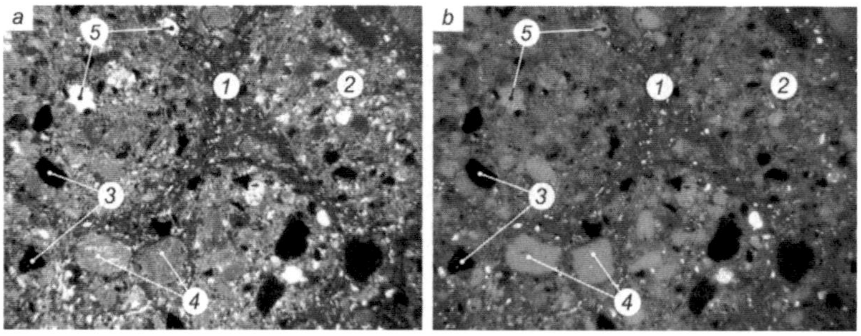

Figure 4. Micrographs of the ceramic samples matrix structure containing waste coal of secondary processing from coal processing plant "Abashevskaya". Thin section, transmitted light, 25×, Nicols II (a); Nicols + (b): 1 – the boundary layer of granules; 2 – granule body; 3 – non-transparent segregations of brown and dark color; 4 – melted silica particles; 5 – pores.

Examination of the ceramic crock form granulated waste coal by the scanning electron microscope also revealed differences in the material structure over the surface of granules contact. At a greater magnification (Figure 5) the melted matrix microstructure containing closed, circular or oval pores (1-10 μm) compared with the disperse phase can be seen, which is consistent with the data of optical microscopy. Besides, in the boundary layer there are macropores of irregularly form, elongated along the granules perimeter.

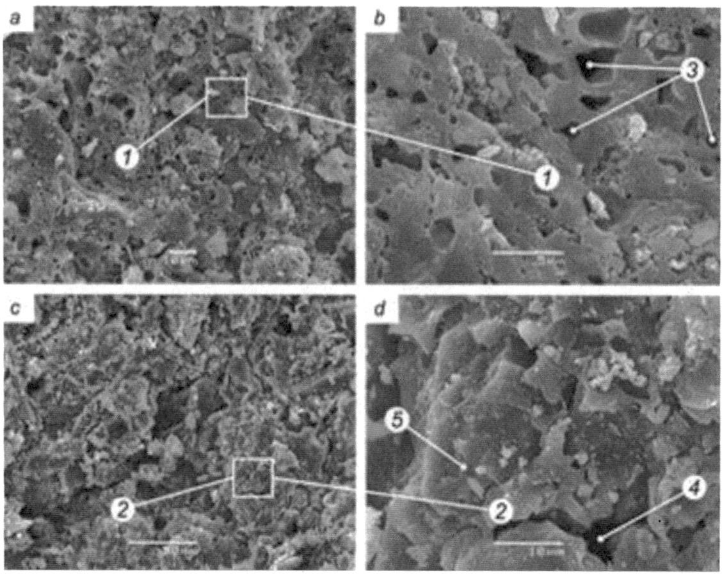

Figure 5. SEM micrograph of dispersion medium areas (a, b) and disperse phase (c, d) of the ceramic matrix composite containing waste coal of secondary processing from "Abashevskaya" plant: 1 – boundary layer; 2 – body of a granule; 3, 4 – closed pores of round, oval and irregular shapes; 5 – melted grains of feldspar.

Thus, in case of use of fine dispersed waste coal as an aggregated filler and actively sintering clay component as a material forming the matrix, we obtain the matrix structure of a ceramic crock, which is a system of nuclei of non-caked material and the dense coked shell, providing high strength and performance characteristics of wall ceramics

5. Conclusions

The studies conducted on the example of waste coal from coal processing plant "Abashevskaya" field proved the possibility of a complex use of waste coal as the main raw material (up to 80 wt.% in the mixture composition) for the production of ceramic bricks after their secondary processing and mixture preparation by the developed method. High physico-mechanical characteristics of the product are achieved due:

- prior fine grinding to class -300 μm of carbonized, coarse technogenic raw material to achieve high uniformity of the molding press-powder and eliminate the harmful effects of carbonate inclusions;
- aggregation of waste coal fine powders into granules to achieve their greater compaction rate and exclusion of air pressing during molding;
- rational particle size distribution of molding mixtures with a predominant granule size 1-3 mm, that allows a uniform structure of a capillary-porous body to be obtained.

6. Acknowledgements

The research was carried out within the state task of the Ministry of Education and Science of the Russian Federation (topic No. 2555 "Development of fundamentals of cost-effective utilization of carbon-containing wastes on the basis of their utilization as fuel for automated thermal power units and production of construction materials from ash residues").

References

[1] Chernyshov E M 2010 *Proc. of Int. Conf. on Achievements and Problems of Materials Science and Construction Industry Modernization* (Kazan: KSUAE) vol 1 pp 8–9.
[2] Shchadov V M 2007 *Coal* **8** 28 –31
[3] Storozhenko G I et al 2013 *Construction Materials* **4** 57–61
[4] Ashmarin G D et al 2010 *Construction Materials* **4** 24–27
[5] Stolboushkin A Yu et al 2011 *Construction Materials* **4** 43–46
[6] Vylomov F A 2015 *Construction* **4** 40–42
[7] Stolboushkin A Yu, Storozhenko G I, Ivanov A I, Berdov G I and Stolboushkina O A 2013 *The Raw Material Mixture for the Production of Wall Ceramics and Method of its Production* pat. of the Russian Federation No. 2500647
[8] Skramtaeva B G 1962 *Reference book on Production of Structural Ceramics* (M.: Srtoyizdat) vol 3 p 608
[9] Kaynarsky I S et al 1948
[10] *Refractory Materials* (M.: Srtoyizdat) p 361
[11] Ivanov A I et al 2015 *Construction Materials* **4** 65–69
[12] Stolboushkin A Yu et al 2011 *Wall Ceramic Products Based on Waste Coal and Iron-ore Containing Additives* (Novokuznetsk: Inter-Kuzbass) p 156
[13] Stolboushkin A Yu 2011 *Construction Materials* **2** 10–15
[14] Stolboushkin A Yu et al 2012 *Construction Materials* **5** 85–89

A study on sintering of ceramic bricks made from waste coal

A Yu Stolboushkin[1], A I Ivanov[1], V A Syromyasov[2] and O A Fomina[2]

[1]Institute of Thermophysics n.a. S.S. Kutateladze of Siberian Brunch of the Russian Academy of Sciences, 1 Akademik Lavrentyev Ave, Novosibirsk, Russia, 630090
[2]Siberian State Industrial University, 42 Kirova Street, Novokuznetsk, Russia, 654007

E-mail: ivanovaliv1989@gmail.com

Abstract. The results of the study on phase transformations and structure formation during firing of bricks made from wastes of coaly argillite processing. The effect of firing temperature on the processes of mineral formation and changes in the quantitative content of amorphous phase and brick porosity was defined. It was found that at 950-1050 °C the sintering takes place providing a solid brick structure.

1. Introduction

Wastes generated by coal extraction and processing can be utilized as a raw material in the production of building materials including bricks but in Russia they are hardly used worsen the environment in the coal-mining regions. Wastes of Korkino lignite open-cut in Chelyabinsk region, Russia, are represented by argillites, aleurolites, coaly argillites, sandstones and shales [1]. Production of ceramic bricks of high quality made of waste coal is possible on the basis of the method developed by the authors [2]. This method of manufacturing ceramic matrix composites allows various types of technogenic and natural materials with low plasticity to be utilized in the building ceramics [3].

Firing of building ceramics is one of the most important factors that determines the properties of the product [4]. In the production of ceramic materials drying and firing consume the greater part of energy resources. Rational modes of heat treatment allows a brick with the required physico-mechanical properties at low cost to be obtained [5, 6].

The aim of the present research was to determine the firing temperature and study the processes of brick sintering made from processing of coaly argillites.

2. Methods and materials

At the first stage to determine the optimum firing temperature for bricks, made of coaly argillites with corrective additive of loam, we prepared a number of laboratory samples.

X-ray fluorescence (XRF) of coaly argillites and clay loam was performed with the use of spectrometer Shimadzu XRF 1800. The chemical composition is given in Table 1.

Table 1. Chemical composition of raw materials.

Raw material	Oxide content in % of dry substance								
	SiO_2	TiO_2	Al_2O_3	Fe_2O_3	MnO	MgO	CaO	R_2O	LOI
Processing wastes of coaly argillites from Korkino deposit	43.85	1.03	16.9	9.97	0.2	2.8	3.35	2.7	19.2
Loam fromNovokuznetsk	59.9	0.9	15.3	5.7	0.2	2.4	5.4	4.8	5.4

The study of the phase composition according to the data of the X-ray diffraction (XRD) showed that the coaly argillites consist of quartz, kaolinite, chlorite, siderite, and muscovite. It also includes dolomite, hematite and feldspar (Figure 1). Mineralogical composition of Novokuznetsk loam corresponds to the group of polymineral clays of kaolinite-montmorillonite-hydromica type. XRD of coaly argillites and clay loam was carried out using spectrometer Shimadzu LabX XRD-6000.

XRD data confirmed the results of the thermal analysis (Figure 2) done on the derivatograph Setaram Lab Sys Evo.

Loam is medium plastic (11.5), low disperse with low content of large and medium-sized inclusions, refers to low-melting, low temperature, non-coking raw material.

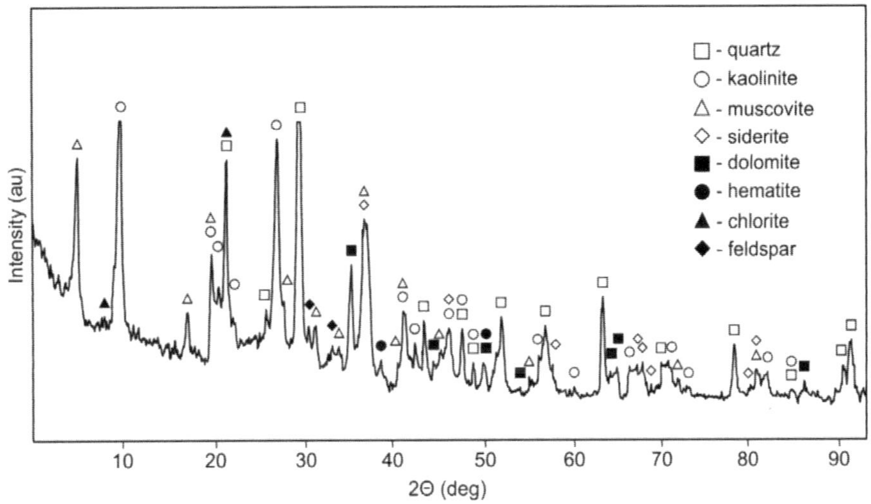

Figure 1. XRD of processing wastes of coaly argillites.

Figure 2 shows the TG, DTA and DTG data of processing wastes of coaly argillites. The mass loss is as follows:100-120 °C dehydration of free water; 470-560 °C dehydration of the clay mineral; 740-780 °C decomposition of carbonates.

Total mass losses after the heating up to 1000 °C amounted to 22.64%, which indicates a high content of carbon dispersed mass.

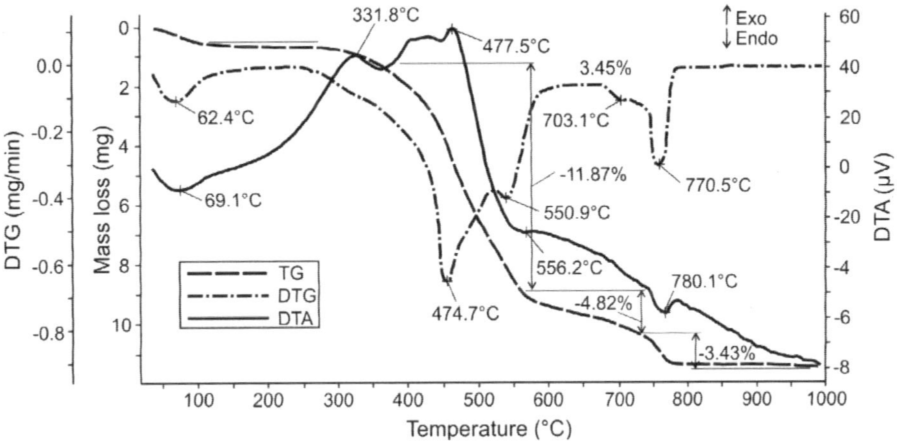

Figure 2. TG, DTA and DTG curves of processing wastes of coaly argillites.

The laboratory ceramic samples from processing wastes of coaly argillites with an additive of loam were prepared by the patented method [2]. The mixture preparation included drying of raw material up to 2 ÷ 3% and grinding to class -300 μm. Particle size distribution after grinding was analyzed using laser diffraction granulometer Malvern Mastersizer 2000. The results showed that the content of fractions was (%): less than 0005 mm – 28%, 0.05÷0.005 mm – 50%, 1÷0.05 mm – 22 %.

The grinded argillites belong to low plastic materials (plastic index ≈ 6.0); non-caking, non-sensitive to drying (the coefficient of drying sensitivity <1) [7].

From the dried and grinded raw materials in the intensive mixer we prepared the molding powders of the following composition (%): processing wastes of coaly argillites 70; loam from Novokuznetsk 30. From the granulated powder at 20 MPa of compaction pressure the samples of cylindrical form 50 mm in diameter and height 45 ÷ 50 mm were made. Firing was conducted at various temperatures from 600 to 1200 °C.

3. Results and discussions

XRD of samples (Fig. 3) fired at a maximum temperature from 600 to 900 °C shows a sharp decrease in chlorite and minerals of carbonate group at a temperature 800 °C, which completely disappear above 900 °C. With the temperature rise starting at 900 °C the content of mineral phases of hematite and augite increases sharply. This indicates the hematitization processes and is confirmed by the characteristic red color of the samples.

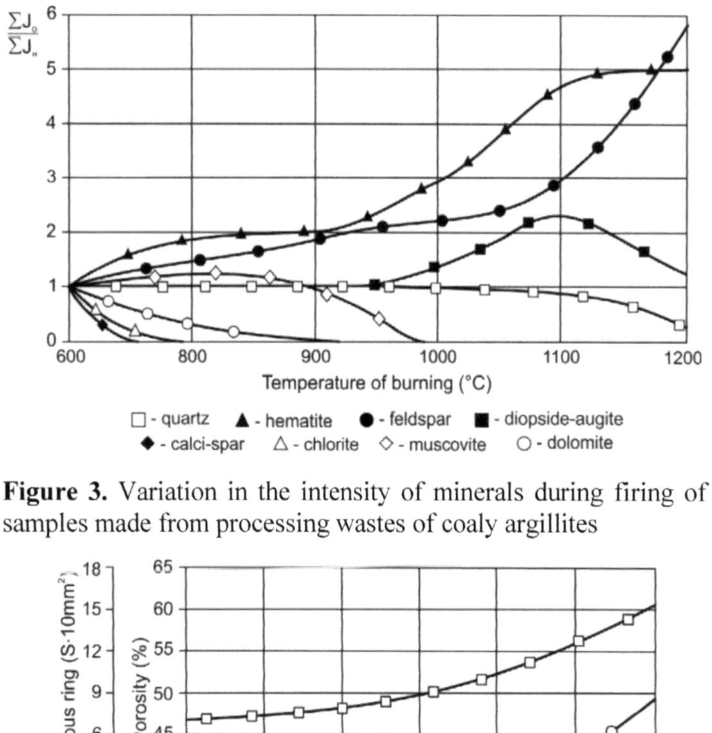

Figure 3. Variation in the intensity of minerals during firing of samples made from processing wastes of coaly argillites

The dependence of the porosity changes in the samples on the firing temperature is given in Figure 4. At a temperature from 600 to 900 °C the total porosity in the samples increases in average by 10-14% due to activation of gas generation in the material. At a temperature 950-1050 °C a slight decrease 1-3% in the porosity occurs, which can be explained by the increase in the amount of pyroplastic phase and partial removal of gases from the material during sintering. Starting from the temperature of 1100 °C there is a melting and swelling of products. A significant increase in the glass phase is evidence of the process of liquid and solid phase sintering of samples [8]. Sintering and formation of a solid structure of ceramic material occur at 950-1050 °C; it is confirmed by high physico-mechanical properties (Table 2).

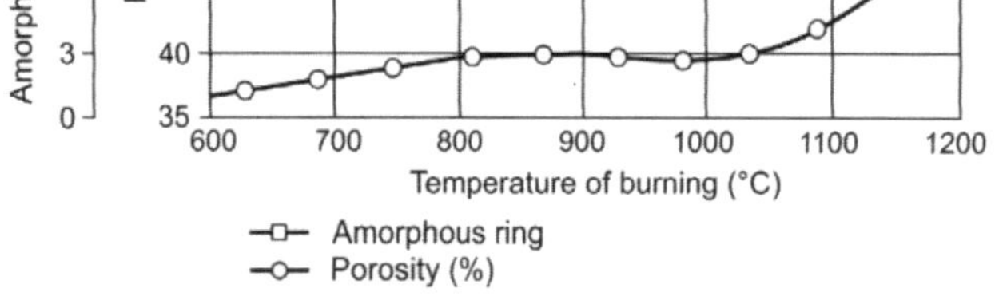

Figure 4. Variation in the content of X-ray amorphous phase and porosity of samples made from processing wastes of coaly argillites depending on the firing temperature.

Table 2. Physico-mechanical properties of ceramic samples.

No.	Firing temperature, °C	Bulk density, kg/m^3	W, %	Ytot, %	Rc, MPa	Strength-density ratio
1	600	1771	18.0	1.3	6.0	3.4
2	800	1684	20.6	2.0	14.0	8.3
3	900	1683	19.9	2.7	21.4	12.7
4	950	1675	19.6	3.0	23.1	13.8
5	1000	1725	16.7	5.4	24.7	14.3
6	1050	1725	14.9	5.7	25.4	14.7
7	1100	-	11.8	-	-	-

On the basis of research at the brick factory JSC "Krasny Kirpich" (Russia, Krasnoyarsk) a pilot batch of bricks made from processing wastes of coaly argillites was produced.

The study of ceramic brick macrostructure (Figure 5) revealed the presence of a continuous ordered frame with the directed orientation of cells that can be attributed to the class of ceramic matrix composites. The matrix of the composite material is formed from a clay component of the mixture, and the aggregate – from processing wastes of coaly argillites.

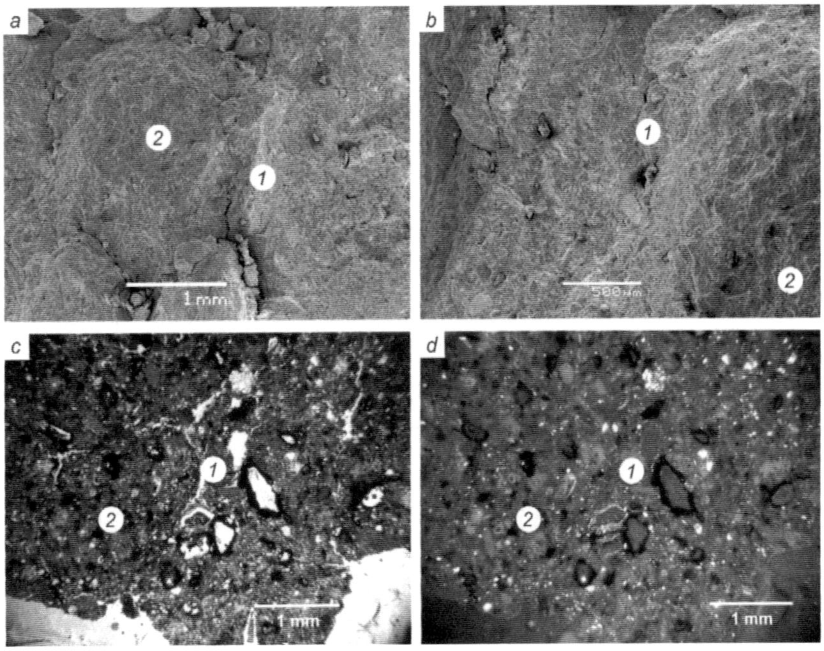

Figure 5.SEM (a, b) and optical microscopy (c, d) of ceramic brick made from processing wastes of coaly argillites: 1 – the boundary layer between granules; 2 – aggregate.

During investigation of thin sections (Figure 6) under the polarizing microscope we observed ore minerals of spherical, sometimes oval or polygonal shapes, caused by dehydration, recrystallization and dissociation in the process of sintering. Partial melting of ore minerals confirmed by isometric form of brown iron crystals. Large fragments of initial minerals are covered with brownish shell, the spaces between them filled with cryptocrystalline and amorphous substance (Figure 5c, d; Figure 6).

Study of the microstructure by the polarizing microscope revealed that after firing the aggregate has a thin brecciated texture and consists of relic minerals and neoformations of crystalline phases connected by the glass phase into a single unit (Figure 6). The boundary layer of the composite matrix is formed from clay minerals in the process of sintering, is a cryptocrystalline aggregate continuously transiting from one granule into another.

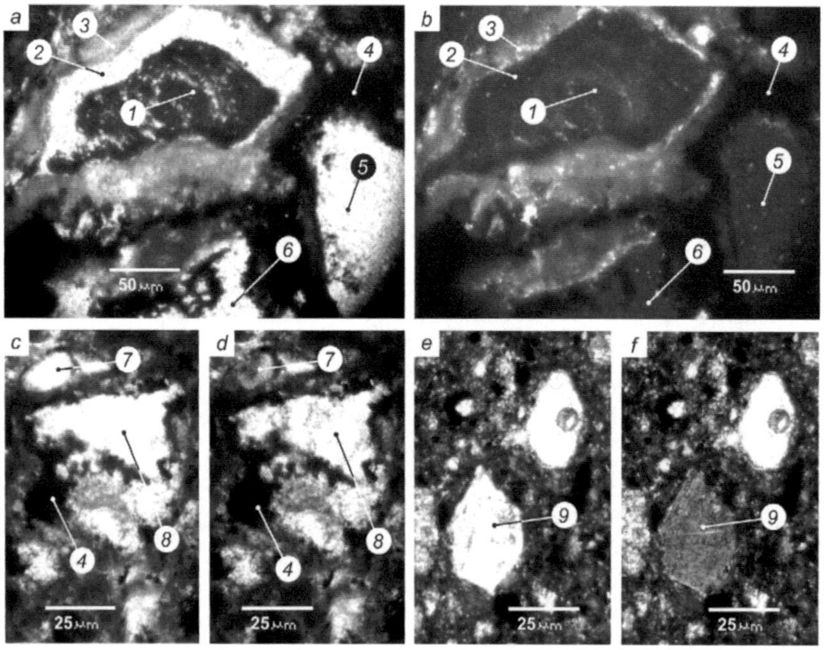

Figure 6. Optical microscopy of the boundary layer of ceramic matrix composite: 1 – neoformations of hematite nature; 2 – pore edge; 3 – glass crystalline phase; 4 – ore minerals; 5 – complex pyroxene; 6 – pore; 7 – augite; 8 – quartz; 9 – rhombic pyroxene.

XRD showed in the aggregate quartz, hematite, feldspar, augite and a mullite-like phase; in the boundary layer: quartz, hematite, mullite, feldspar and spinel (Figure 7).

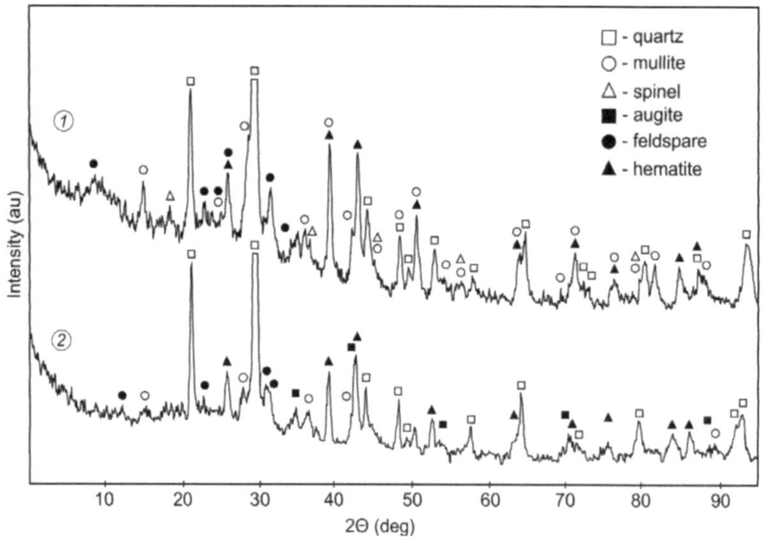

Figure 7. XRD of ceramic bricks made from processing waste of coaly argillites: 1 – boundary layer; 2 – aggregate.

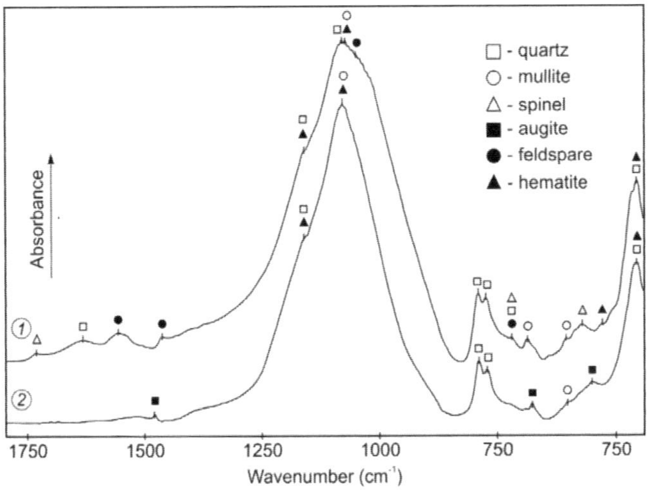

Figure 8. IR-spectra of ceramic bricks made from processing waste of coaly argillites: 1 – boundary layer; 2 – aggregate.

By the pronounced x-ray amorphous halo we can judge about a significant amount of glasses in the dispersion medium (15-20%), which is consistent with the observation in thin sections. Particles of relic quartz and feldspar alongside with the minerals of neoformations are reinforcing matrix, having a fine disperse hematite nature. Infrared spectra of the ceramic matrix composite confirm the presence of the indicated mineral phases (Figure 8).

The increased pore content in the granule body is due to burn-out of disperse coal in the process of firing. In the boundary layer the porosity slightly increases due to air pressing during granulated molding mixture compression, as well as squeezing water into the dusting layer during adobe brick molding. Walls of pores in the boundary layer are formed by an amorphous glassy mass permeated by cryptocrystalline minerals.

6. Conclusions

As a result of the conducted research into process of sintering during firing bricks from processing waste of coaly argillites it was established found:

- content of main mineral phases remains almost unchanged during firing up to 1100 °C, which positively affects the structural and mechanical characteristics of the bricks;
- increase in the firing temperature above 1100 °C leads to overfiring of bricks;
- optimum amount of pyroplastic glass phase for brick sintering is formed at a temperature 950-1050 °C;
- high physico- mechanical properties of bricks are conditioned by the spatially organized matrix structure.

7. Acknowledgements

The research is performed at the Institute of Thermophysics n.a. S.S. Kutateladze of Siberian Brunch of the Russian Academy of Sciences, with support of the Ministry of Education and Science of the Russian Federation, Agreement No. 14.607.21.0106, unique ID number RFMEF–160714 X 0106.

References

[1] Nikolayenko M A et al 2008 *Proc. of the RAACS Meeting on Health of the Population - Living Environment Development Strategy* (Belgorod) pp 377–381

[2] Stolboushkin A Yu, Storozhenko G I, Ivanov A I, Berdov G I and Stolboushkina O A 2013 *The Raw Material Mixture for the Production of Wall Ceramics and Method of its Production* pat. of the Russian Federation No. 2500647

[3] Stolboushkin A Yu 2011 *Construction Materials* **2** 10–13

[4] Pavlov V F 1976 *Physico-chemical Basis of Building Ceramics Products Firing* (M.: Stroyizdat) p 240

[5] Kochneva T P 2003 *Construction Materials* **2** 39–41

[6] Kotlyar V D et al 2013 *Construction Materials* **4** 44–48

[7] Tkachev A G et al 2013 *Technique and Technology of Silicates* **2** 17–21

[8] Stolboushkin A Yu et al 2013 *Construction* **2–3** 28–36

IOP Conf. Series: Earth and Environmental Science **45** (2016) 012019 doi:10.1088/1755-1315/45/1/012019

Structure formation of aerated concrete containing waste coal combustion products generated in the thermal vortex power units

A I Ivanov, A Yu Stolboushkin, M V Temlyanstev, V A Syromyasov and O A Fomina

Siberian State Industrial University, 42 Kirova Street, Novokuznetsk, 654007, Russia

E-mail: ivanovaliv1989@gmail.com

Abstract. The results of fly ash research, generated in the process of waste coal combustion in the thermal vortex power units and used as an aggregate in aerated concrete, are provided. It is established that fly ash can be used in the production of cement or concrete with low loss on ignition (LOI). The permitted value of LOI in fly ash, affecting the structure formation and operational properties of aerated concrete, are defined. During non-autoclaved hardening of aerated concrete with fly ash aggregate and LOI not higher than 2%, the formation of acicular crystals of ettringite, reinforcing interporous partitions, takes place.

1. Introduction

Since the second half of the XX century the coal mining has significantly increased, year on year volumes of industrial production are rising and power generation facilities are being rapidly developed. It all results in generation of billion tonnes different types of waste. The most dominant types of industrial wastes are ashes and sludges from thermal power plants (TPP). Just two countries, China and India, account for 85 percent of all new coal capacity, which corresponds to the launch of two large TPP in a week [1].

For the effective management of industrial wastes all over the world the specialized organizations are founded, for example, by member states of the European Coal Combustion Products Association (ECOBA) 90% of ash is processed [2]. China recognizes the importance and urgency of addressing the environmental problems within the framework of government programs. There the wastes from coal mining and heating energy sector are widely utilized as well as other technogenic wastes [3].

In Russia annually more than 3 billion tonnes of industrial wastes are produced [4], however, their re-use remains insignificant; in particular, the amount of wastes utilized in the heating energy sector is less than 15% of the volume of their generation, though the potential of wastes use for the manufacture of building materials varies from ash binder to ceramic bricks [5].

The aim of the investigation was to study the material composition and evaluate the ashes, produced by coal-containing wastes combustion in the thermal vortex power units, to be further used as a raw material for non-autoclaved aerated concrete production.

2. Methods and materials

During fuels combustion in the thermal units the carbon content in the ash depends on a number of factors, the main of which are the combustion technology and design features of the thermal power

Figure 1. Automated thermal power unit of vortex type in the laboratory of Siberian State Industrial University.

unit [6]. The design of pulse vortex furnace (Figure 1) developed for combustion of dry powdered refuse coal and other types of ballasted solid fuel (waste coal, sludges, shales, lignites) [7] influences greatly on the material composition of ashes examined in the present paper.

At the first stage we carried out the comprehensive study of the material composition of ashes obtained after waste coal combustion. Before taking samples various combustion modes of vortex chamber were set, simulating the real operation conditions of industrial thermal power plants, to get ashes with different carbon content. Ash residues from fuel combustion were conditionally divided into three samples according to carbon content (%): <1.0; 1.0-3.5; 3.5-7.5.

X-ray fluorescence (XRF) was done for three samples of fly ash using spectrometer Shimadzu XRF 1800. The chemical composition is given in Table 1.

Table 1. Chemical composition of fly ash samples from combusted waste coal.

Sample	Mass fraction of the components (%)											
	SiO_2	TiO_2	Al_2O_3	CaO	MgO	Fe_2O_3	R_2O	MnO	S	P	C	LOI
1	53.38	0.95	21.52	7.50	2.18	7.66	5.39	0.18	0.88	0.24	<0.1	0.50
2	53.51	0.89	17.89	6.73	1.85	8.47	4.37	0.09	0.53	0.13	2.09	3.46
3	44.09	1.40	18.65	5.60	1.82	10.84	4.22	0.12	1.12	0.20	4.34	7.60

The study of the phase composition according to the data of the X-ray diffraction (XRD) showed that the fly ash consists of quartz, hematite, field spar and muscovite. There are also calci-spar and anhydrite (Figure 2). Halo on the diffraction pattern shows a significant quantity of amorphous phase. XRD was done on one average fly ash sample using spectrometer Shimadzu LabX XRD-6000.

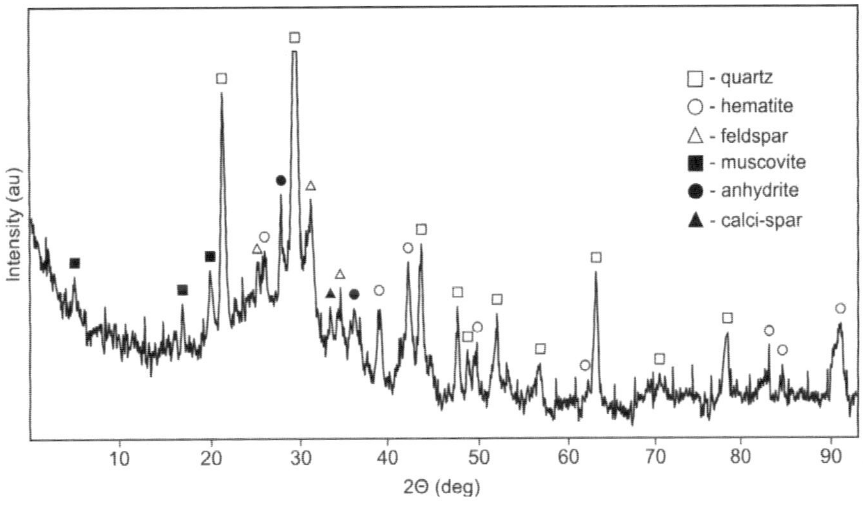

Figure 2. XRD analysis of the average fly ash sample from combusted waste coal.

Figure 3 shows the TG and DTA data of the average sample of fly ash from combustion of coal waste. Within the temperature range from 400 to 1000°C there is a change in the mass on the TG curve, which corresponds to carbon combustion. This process appears on the DTA curve as exothermic with maximum at 556°C. Endothermic peak is on the DTA at 807°C due to the decomposition of carbonates and shows its small quantity.

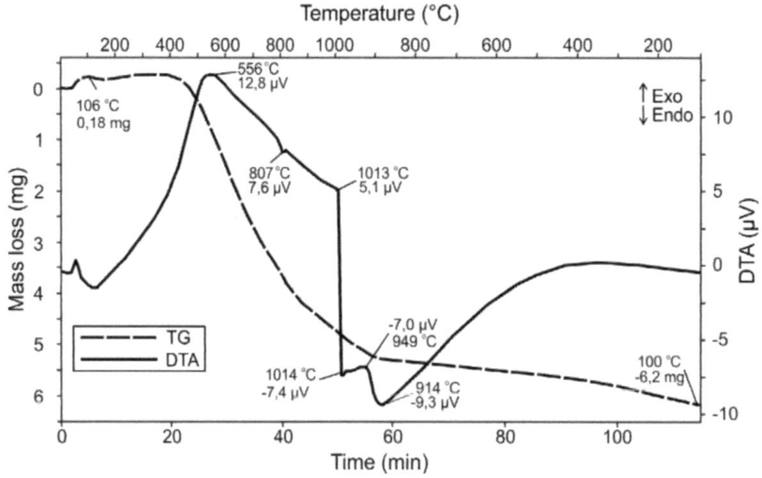

Figure 3. TG and DTA curves of the average fly ash sample from combusted waste coal.

The LOI of samples is approximately 5%; this result corresponds to category A of BS EN 450-1:2012 [8]. But the other sample shows LOI of approximately 7%, which is too high to be used in cement or concrete production. Particle size distribution analysis was done for fly ash samples using laser diffraction granulometer Malvern Mastersizer 2000 operating a Hydro 2000G module (Table 2).

Table 2. Specific surface area and particle size of fly ash samples from combusted waste coal.

Sample	Bulk density, kg/m^3	SSA, m^2/kg	Particle size in %				
			>200μm	200-80 μm	80-60 μm	60-40 μm	<40μm
1	950	291	0.15	3.63	2.81	3.21	90.2
2	980	278	0.91	7.73	6.79	7.63	76.93
3	1025	270	1.94	14.16	12.81	14.34	56.75

Thus, the ash from waste coal combustion in the thermal vortex power unit can be used as a disperse agglomerate for production of building materials with a conglomerate or mesh structure.

3. Results and discussion

At the second stage of the study using the standard method we carried out the selection of composition of aerated concretemixture on the base of the studied ash. For 1 m^3 of aerated concrete with bulk density 600 kg/m^3 the content of component was:

- CEM III/A 42.5N (75 % in the mixed binder) 190.5 kg;
- lime (25 % in the mixed binder) 70.6 kg;
- ash 317.5 kg;
- water 343.0 l;
- aluminum powder 0.39 kg.

To prepare standard samples with sizes 70×70×70 mm we used ash with different LOI (from 0.1 to 7.6%) as an aggregate. In the course of the experiment the volume increase ratio of gas concrete mixture was determined. It was found that the samples containing ash with a high LOI the volume increase ratio was 2.6 times lower compared with the samples with almost no LOI. These results can be explained by the interaction between the residual carbon of ash component and a part of the calcium hydroxide of lime component. As a result, the estimated mass balance between the components, involved into the gas formation reaction, was broken, which led to the increase of bulk density.

The samples were subjected to the natural hardening for 28 days at a temperature 20-22 °C. The test results of physico-mechanical and thermal properties of samples are presented in Table 3. The structure of non-autoclaved aerated concrete from fly ash with low LOI is in Figure 4 (a, c, e) and high LOI – Figure 4 (b, d, f).

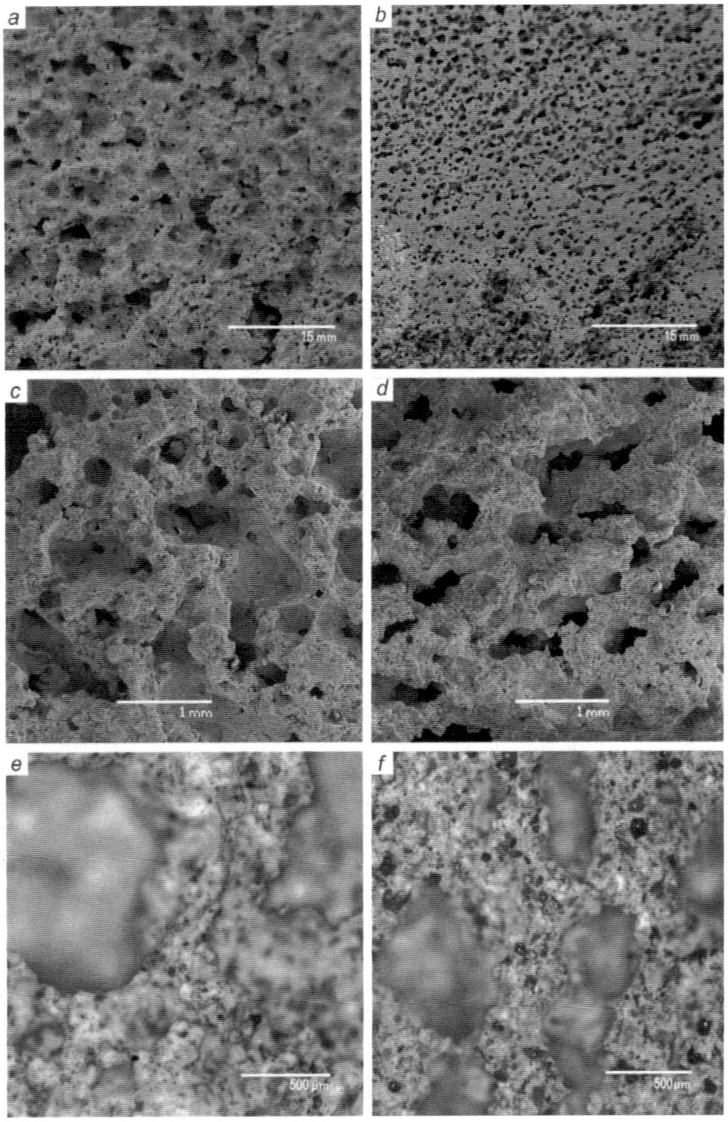

Figure 4. Macrostructure (a, b), SEM (c, d) and optical microscopy (e, f) of aired concrete based on the fly ash from combusted waste coal.

Table 3. Physico-mechanical properties of non-autoclaved aerated concrete from fly ash with different LOI.

Indicator	LOI (%)		
	<0.1	3.46	7.60
Bulk density, kg/m^3	548	596	1486
Open porosity, %	44.7	43.3	38.9
Compressive strength, MPa	5.7	4.9	2.6
Strength-density ratio	10.5	8.5	3.5
Thermal conductivity, W/(m.K)	0.105	0.120	0.470

Specifics in the structure of solid phase interporous partitions largely determine the performance of aired concrete and depend on the sizes, shapes, spatial arrangement and interaction of individual components of the mineral phases [9, 10].

According to XRD data in the aerated concrete (after 28 days) mineral phases are established: quartz, hydrosilicates CSH(B) and C_2SH_2, ettringite. These results are confirmed by the thermal analysis of the material.

SEM confirmed the formation of hydrosilicates in the samples. When using ash with low LOI, the concrete structure is permeated with needle-like crystals of ettringite, providing the micro-reinforcement of interporous partitions (Figure 5a, b). In contrast, in the ash samples with high LOI the ettringite crystals are in the embryo forms, and the aerated concrete has a dense texture (Figure 5c, d).

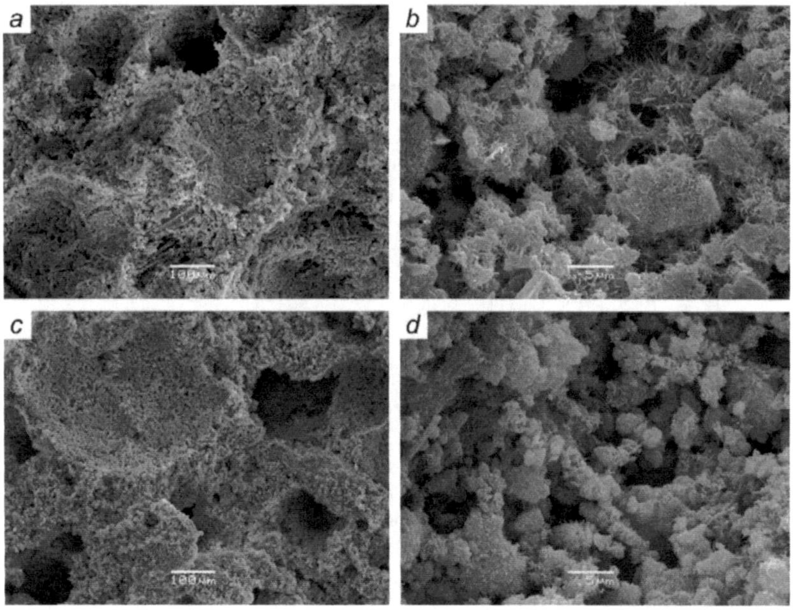

Figure 5. SEM of non-autoclaved aerated concrete from fly ash with low (a, b) and high (c, d) LOI.

5. Conclusions

As a result of the research the following conclusions can be drawn:

- use of ash from waste coal combustion in the thermal power units of a vortex type provides high performance properties of aired concrete products with an bulk density less than 600 kg/m^3 at low (<4%) LOI in the ash aggregate;
- LOI greater than 4% prevents the formation of cellular concrete structure due to disturbances of the gas formation process in the mixture, which leads to the increase in the bulk density of

the material (2.5-2.7 times) and increase in thermal conductivity coefficient up to 0.5 W/(m.K);
- during hardening of aired concrete mixture with ash aggregate (LOI less than 2%) the formation of acicular crystals of ettringite reinforcing interporous partitions occur, thus, strength of the aired concrete increases;
- use of ashes with LOI less than 3.5 % allows the aerated concrete with the bulk density 550-600 kg/m^3, compressive strength 5-6 MPa and thermal conductivity coefficient 0.105-0.120 W/(m.K) to be produced.

6. Acknowledgements
The research was carried out within the state task of the Ministry of Education and Science of the Russian Federation (topic No. 2555 "Development of fundamentals of cost-effective utilization of carbon-containing wastes on the basis of their utilization as fuel for automated thermal power units and production of construction materials from ash residues").

References
[1] Shearer C et al 2016 *Tracking the Global Coal Plant Pipeline. Report March*http://www.sierraclub.org/files/uploadswysiwig/final%20boom%20and%20bust%202017%20(3-27-16).pdf
[2] Feuerborn HJ 2011 *World of Coal Ash (WOCA) Conference* (Denver) vol 1 pp 125–148
[3] *Law for Promotion of Cleaner Production of PRC* 2002 http://www.npc.gov.cn/englishnpc/Law/2007-12/06/content_1382101.htm
[4] Khoroshavin L B et al 2016 *Basic Technologies for Processing of Industrial and Municipal Solid Waste* (Ekaterinburg: UrFU) p 219
[5] Christy C F et al 2011 *Asian J. of Civil Eng. (Building and Housing)* **12** (1) 87–105
[6] Bagryantsev V I et al 2013 *Bulletin of SibSIU* **3** 33–38
[7] Bagryantsev V I et al 2013 *Bulletin of SibSIU* **4** 36–41
[8] *BS EN 450-1:2012.* Fly Ash for Concrete. Definition, Specifications and Conformity Criteria
[9] Mysatov I A 1971 *Study of Basic Regularities of Formation of Macrostructure in Large Arrays of Aerated Concrete* PhD thesis (Leningrad) p 165
[10] Silaenkov E S 1986 *Durability of Goods from Cellular Concrete* (M.: Stroyizdat) p 176